Economics - Business/Accountancy.

Business Calc

Other titles in this series

Business Communication
Business Law
Business Studies
The Business of Government

Business Calculations

David Browning BSc, C.Chem, FRSC, ARTCS, is a well-known author of mathematics and science textbooks. He is a mathematics examiner of long standing. Until recently he was Head of Chemistry and Food Technology at Bristol Polytechnic. He is now a full-time author and education consultant.

Business Calculations

David Browning

Chambers Commerce Series

© David Browning 1987

Published by W & R Chambers Ltd Edinburgh, 1987

All rights reserved. No part of this publication may be reproduced or transmitted in any form or by any means, electronic or mechanical, including photocopying, recording or any information storage or retrieval system without prior permission, in writing, from the publisher.

Every effort has been made to trace copyright holders, but if any have inadvertently been overlooked the publishers will be pleased to make the necessary arrangements at the first opportunity.

British Library Cataloguing in Publication Data

Browning, David
    Business calculations.—(Chambers commerce series)
    1. Business mathematics
    I. Title
    513'.93   HF5691

    ISBN 0-550-20700-7

Typeset by Waddie & Co. Ltd Edinburgh

Printed in Great Britain by
Richard Clay Ltd, Bungay, Suffolk

Contents

Preface   vii

Chapter 1   The Needs of Business

1.1   The Profit Motive   1
1.2   Related Numerical Skills   2
1.3   The Role of the Calculator and Computer   3
1.4   Types of Problems   4
      Exercise 1   6

Chapter 2   Basic Arithmetic Operations

2.1   The Process of Addition   8
2.2   Subtraction   13
2.3   Multiplication   18
2.4   Division   21
2.5   Combinations of the Four Rules   25
2.6   Calculator Notes   27
      Exercise 2   28

Chapter 3   Fractions

3.1   Vulgar Fractions   30
3.2   Decimal Fractions   45
3.3   Approximations   54
3.4   Calculator Notes   57
      Exercise 3   59

Chapter 4   Metric Measurements

4.1   Measurement of Length   60
4.2   Measurement of Capacity (Volume)   62
4.3   Measurement of Weight   65
      Exercise 4   67

Chapter 5 Ratio, Proportion and Percentages

5.1	Ratios	68
5.2	Proportion	69
5.3	Percentages	77
5.4	Interests and Discounts	84
	Exercise 5	101

Chapter 6 Areas and Volumes

6.1	Measurement of Area	104
6.2	Measurement of Perimeter Length	108
6.3	Measurement of Volume	110
	Exercise 6	111

Chapter 7 Visual Presentation of Information

7.1	Tables	112
7.2	Graphs and Charts	124
7.3	Straight Line Graphs	137
	Exercise 7	142

Chapter 8 Averages

8.1	Averages	147
	Exercise 8	153

Chapter 9 Statistics and the Presentation of Business Data

9.1	Tabular Presentation	155
9.2	Pictorial Presentation	163
	Exercise 9	175

Chapter 10 Assignments	179
Answers	187
Appendix 1	206
Appendix 2	209
Index	211

Preface

Business Calculations is designed to take students through the basic essentials of arithmetic, graphwork and statistics, and to provide them with a sound framework of knowledge for courses up to and around GCSE level. This includes the elementary courses in business calculations of BTEC and the professional associations.

There has been a serious attempt at every stage to relate the calculations to real life business situations and to show quite clearly the relevance of all the material included in the book. Examples of the use of calculators have been included and sufficient information has been given to allow students to modify the programs to suit their individual calculators.

The content has been developed slowly and thoroughly to enable readers to study and understand the ideas of business mathematics fully. Many worked examples are included. In addition the answers to the questions contain hints where necessary on how to attempt the calculations.

D. B.

Chapter 1

The Needs of Business

What is business? Why study business calculations when in the age of the calculator and the computer all our calculations can be done for us?

In this chapter we will look at these questions in the light of what businesses need.

To start with we must define what a business is. The word business is used in many ways. 'A trade, profession or occupation' is as good a description as any. This can vary from a freelance consultant to a multinational company. In many ways the calculations involved are fundamentally the same. However, the scale and complexity of operations are very different.

All of us are involved in using business calculations to some extent. We have to pay for goods and services provided. We must pay our taxes. We need to understand our bank balances and so on. Therefore we need to know how business people make up their bills. We also need to be able to check if they are right.

Ask a businessperson what his or her main object is in running, or working for, a business. He or she will admit, if they are honest, that it is to make money. We need money to live. The creation of wealth, if reasonably distributed, creates an affluent society. Let us examine the profit motive more closely.

1.1 The Profit Motive

It is the objective of all businesses to make money and to be both efficient and cost-effective. This involves many skills. These include sound pricing strategies, good stock control, accurate and efficient secretarial facilities. All of these are important. If we look carefully at the organisation of any company, no matter what size, it is easy to see that much of the work, even in financial matters, involves an exchange of information only. For example, if you order a product from a firm you are likely to pay by cheque. This entails presenting a piece of paper to the company who will pay it into their account. The value on the cheque will be credited to their bank account. No actual money changes hands. A consultant may be employed by a firm to assess the problems occurring in an engineering process. He will eventually produce a report on his findings. This again is an exchange of information. Many other examples may be found.

However there is sufficient evidence to show that
(a) efficient information exchange and
(b) a reliable and cost-effective provision of services and products
are likely to make a company profitable. The collection, interpretation and use of quantifiable data are essential for success. Therefore we must be able to understand the mathematical processes involved.

One point needs to be mentioned. Businesses are not run in isolation but nearly always in competition with other businesses. This factor is important in all profit making considerations.

1.2 Related Numerical Skills

An understanding of the four basic rules of arithmetic is essential to any study of numerical problems. Everything else is dependent on these. They are the rules of addition, subtraction, multiplication and division. Many basic accounting transactions, such as the payment of bills, depend only on these rules. In most business transactions the numbers involved are rarely whole numbers but are fractions, particularly decimal fractions. Also nowadays we are almost exclusively involved in metric measurements.

One area of business calculations that is of prime importance but which is often a source of difficulty is that of ratios and percentages. These are involved, for example, in determining discount values, mark-up values, tax and insurance rates. They are not restricted to financial considerations but can include personnel deployment and other aspects of business organisation. This traditionally difficult subject will be explained in a simple and readily understood way.

Measurement of areas and volume may seem to be of little relevance but many companies find this knowledge invaluable. In buildings, for example, the shape is very significant. By careful design the same space can be achieved with considerable savings in material and man-hours. On a more mundane plane, careful manipulation of the coverings on, say, chocolate bars can save many thousands of pounds in a year. These kind of manipulations can only be carried out if you understand 'shape' although this is often only required at an elementary level.

Earlier it was mentioned that a major part of a firm's work was in the transmission of information. This can be most immediately perceived and understood when presented in a visual way. This is particularly true of figures. Advertisers use the techniques of charts and graphs very widely and with considerable success. Also technical and commercial presentations to customers or clients can be very compelling using visual methods. Again it is important to understand what you are doing to exploit the techniques successfully.

Last, but not least, is the need to be able to look at past figures and make forecasts for the future. This is achieved by the use of statistics. Statistics is a very large subject and when used for monitoring quality control or quality assurance it demands a great deal of expertise. But even at an elementary level statistics can be a powerful tool if you know how to use it. For example, if you are involved in planning the sales of a service in an expanding town, it is necessary to survey past sales. On the basis of this and other graphical and statistical information an attempt to forecast future sales may be made.

I hope that this shows the significance of the numerical skills you will need. The material is deliberately introduced in such a way that you will be able to follow it on your own. There are enough examples to highlight and show clearly how to use the method you need.

1.3 The Role of the Calculator and Computer

The question most frequently asked about numerical skills is 'Why do we need them? The calculator can do it all for us.' Let us take a very simple example. You have just worked out that your electricity bill is correct using a calculator. In each case you worked out the total bill using + on the calculator. Why did you do this? Because the figures have to be added. How do you know they have to be added? You have to understand what adding means to answer this. When working with more difficult ideas, for example, graphs and statistics, it is even more important to understand what you are doing.

The calculator and the computer are mainly tools to take the monotony and drudgery out of what is often called 'number crunching'. In statistics you often have to add, for example, 50 or more figures and find the average. It is easier to let your calculator do all the adding and division once you have entered the figures but the result is valueless unless you know what an average means.

In business, computers are now used to cut down paper work, and repetitious working out of bills and statements. One good example of this is the so-called cashless society, an extension, in some ways, of the use of credit cards.

4 The Needs of Business

There are many difficulties to be overcome before we have a cashless society. The Truck Act of the last century would need to be repealed. (This gives everyone the right to be paid in cash.) Unfortunately successive governments have avoided taking this decision because it will probably be unpopular. The 'her indoors' syndrome is also still with us. Payment in cash makes it more difficult for a wife (her indoors) to find out what a husband earns. Also many people just feel happier with 'cash in their hand'.

Nevertheless, it is envisaged by many banks that the use of money and cheques will disappear. They will use a method known as electronic funds transfer (EFT). It works like this. You have stayed in a hotel and on leaving you are presented with a bill. You give the receptionist your credit card. This is placed in a machine which links into the computer at your bank, debits your account and credits the amount to the hotel. No paper work is involved apart from your bill. Some supermarkets are already using EFT on a trial basis. This uses the receipt of goods at the till. Presumably you would:

(a) check your bill was added correctly
(b) check your bank statement to see that it tallied correctly.

Again, an understanding of addition is needed.

The examples chosen are very simple ones and once you understand what you are doing the actual addition can be done by machine.

Remember that the machine will only work on the figures you give it. If these are wrong, your answer will be wrong. You would not be happy if the hotel charged you £83.50 instead of £38.50. The machine is only as good as the person using it. Errors can be expensive.

Throughout this book the use of machines will be clearly indicated.

1.4 Types of Problem

In this section two problems will be outlined to show the way in which calculations slot in to business situations.

Problem 1
Some supermarket chains are investigating changing to electronic funds transfer and computerised organisation. Which parts of the organisation can be computerised?

(a) There could be an EFT link between the till and the bank. This not only saves paper work but since bills are automatically paid it cuts down bad debts. Customers will only be able to buy if they are in credit at their bank or if arrangements have been made to be overdrawn.

(b) Items purchased can be listed to the customer and to the stock control department in the store and for the chain. Stock can be automatically recorded. Decisions have to be made on the size of stocks to be maintained in each store and at the chain's central depot. The levels should ensure keeping minimum stocks with maximum turnover. This would enable the chain to maintain the most economical storage space. It would also ensure customer satisfaction.
(c) By timing customers' movements through tills over a set period the pattern of through-put can be determined. This can be used as a basis for future staff deployment giving greater efficiency. It also decreases waiting time for customers.

A statistical study of (b) and (c) will be required for sensible forecasting. This could be done in part by a company statistician and in part by store management staff.

There are other aspects of store management that could be considered but I think a case has been made for a need to understand at least one area of business calculations.

Problem 2
You are a builder with a commission to build a bungalow which is cost-effective but spacious. Clearly you have to work within certain building specifications. However, there are many decisions to be made. These would probably be done in consultation with your client.

(a) What is the best shape for the bungalow? As you will see in Chapter 6 a square building has less outer wall space than a rectangle of the same area. The farther you get from the square, the more costly the walls. However you will have to offset this against the different costs of beams. You may also need more load-bearing walls inside the building to support the beams.
(b) Does the energy saving merit the original cost of loft insulation? How long will it take to get the money back? Are other forms of insulation such as cavity wall insulation or floor insulation cost-effective? Is it possible to assess changes in energy costs in the future?

As in problem 1 only a few aspects of the overall problem have been investigated but these involve the mathematics of Chapters 5 and 6. Both problems also involve the mathematics dealt with in Chapters 2 to 5 to a very great extent.

Working on the basis that, 'if it is useful you will learn it', I hope you are encouraged to read on.

6 The Needs of Business

Exercise 1

The following questions are designed to get you to look at the problems encountered in commerce. Try to find out in each the kind of decisions you will have to make, especially those involving calculation. Then look at the answers on page 187. Try to answer the questions as far as you can without looking anything up, at least to start with.

1. You have decided to set up a small typing agency to produce technical reports for a number of local companies. You obviously want to make a profit. How do you assess local company needs now and in the future? How do you then decide on the number of your staff and the staff structure? What equipment do you need? How are you going to pay for it? Look at these and other questions you consider important. Put down some of the things you will have to calculate.

2. You are opening a small brewery. You have to decide if you want one large boiler or three small boilers in your boiler room. Try to think of the things you would look for in reaching a decision.

3. A company employs 55 men. You are asked to design a new simple salary slip for the company. All salaries are paid directly into the bank. How would you go about this problem?

Now that you have tried the questions look at my suggestions on page 187. Can you see that you will need some knowledge of business calculations to solve the problems?

You will meet similar problems, and others, in Chapter 10. By that stage you will be able to do some of the calculations.

Chapter 2

Basic Arithmetic Operations

There are four basic processes in arithmetic: addition, subtraction, multiplication and division. It is likely that you are familiar with these and know how to use them. However, it is worthwhile ensuring that we understand them fully.

First of all we must look at the way we number things. We use only ten figures to make up all the numbers we need. These are:

$$0, 1, 2, 3, 4, 5, 6, 7, 8, 9$$

The first number above nine is ten which is 10. The numbers then go:

$$10, 11, 12, 13, 14, 15, 16, 17, 18, 19$$

The next number is 20 which is 2 tens. We then get 30 for 3 tens, 40 for 4 tens up to 90 for 9 tens.

So 56 is 5 tens and a 6.

We can put this in a tens column (T) and a units column (U). Note that the order is important.

	(T)	(U)
fifty-six is 5 tens and 6 units	5	6

Put the following numbers in their correct place in the tens (T) and units (U) column below: forty-eight, twenty-six, eighteen.

	(T)	(U)
forty-eight is 4 tens and 8 units	4	8
twenty-six is 2 tens and 6 units	2	6
eighteen is 1 ten and 8 units	1	8

99 (ninety-nine) is the highest number which can be put in two columns. The next number is one hundred and this has three figures 100. The one appears in a new column called a hundreds column (H).

8 Basic Arithmetic Operations

Put the following numbers in their correct place in the hundreds (H) tens (T) and units (U) columns: one hundred and forty-eight, seven hundred and thirty-nine, nine hundred and ninety-nine.

	(H)	(T)	(U)
one hundred and forty-eight is			
1 hundred, 4 tens and 8 units	1	4	8
seven hundred and thirty-nine is			
7 hundreds, 3 tens and 9 units	7	3	9
nine hundred and ninety-nine is			
9 hundreds, 9 tens and 9 units	9	9	9

The number above 999 is one thousand which goes in the next column, i.e. the thousands (Th) column.

	(Th)	(H)	(T)	(U)
one thousand	1	0	0	0

All numbers can be built up in this way. Since this system of numbers involves 10 units it is called the denary or decimal system. Other number systems are possible. For example in computing we use a system called the binary system with only two numbers, 0 and 1. All the numbers we use here are called integers.

2.1 The Process of Addition

The method of collecting things of the same kind together is called addition. If you have a five pound note, a pound coin and a two pound coin, you have five pounds and one pound and two pounds, which is £5 + £1 + £2. You have £8.

You have collected the numbers together or added them. The '+' called 'plus' means the numbers are to be, or have been, added. Other items of the same kind can be added:

e.g. 5 washing-machines + 1 washing-machine + 2 washing-machines gives 8 washing-machines. However, 5 washing-machines + 1 washing-machine + 2 cookers are not 8 washing-machines but are 6 washing-machines + 2 cookers.

You can only add things if they are the same.

Addition can be done in another way:

$$\begin{array}{r} 5 \\ 1 \\ 2 \\ \hline 8 \\ \hline \end{array}$$

Basic Arithmetic Operations 9

When adding larger numbers use the following method:

$$
\begin{array}{cc}
(T) & (U) \\
2 & 1 \\
1 & 4 \\
3 & 3 \\
\hline
\end{array}
$$

Looking at the units column you have:

$$3 + 4 + 1 \text{ is } 8$$

In the tens column you have:

$$3 + 1 + 2 \text{ is } 6$$

$$
\begin{array}{cc}
(T) & (U) \\
2 & 1 \\
1 & 4 \\
3 & 3 \\
\hline
6 & 8 \\
\end{array}
$$

What about:

$$
\begin{array}{cc}
(T) & (U) \\
4 & 5 \\
1 & 4 \\
2 & 7 \\
\hline
\end{array}
$$

Here $5 + 4 + 7$ is 16, that is, 10 and 6.

So you get:

$$
\begin{array}{cc}
(T) & (U) \\
4 & 5 \\
1 & 4 \\
2 & 7 \\
\hline
 & 6 \\
\end{array}
$$

and 10 is left over. In the tens column you have $4 + 1 + 2$, but you also have 1 ten from the units column.

∴ You have $4 + 1 + 2 + 1$ carried over from units. This gives you 8. (The symbol ∴ means therefore.)

Basic Arithmetic Operations

The answer can be written as:

```
    (T) (U)
     4   5
     1   4
     2   7
    ─────
     8   6
         1  ← This is the number carried over.
```

In this example you have added from top to bottom. In the previous one you added from bottom to top. Both ways give the same answer.

Worked examples

Add the following:

(a) 32   (b) 43   (c) 49   (d) 63   (e) 128
     41   28   29   39   136
     16   24   15   44   392
     ──   ──   ──   ──   ───

     ──   ──   ──   ──   ───

Answers

(a) $2 + 1 + 6$ is 9 in the units column.

$3 + 4 + 1$ is 8 in the tens column.

```
       (T) (U)
  ∴     3   2
        4   1
        1   6
       ─────
        8   9
```

(b) $3 + 8 + 4$ is 15.

This is 1 ten and 5 in the units column.

$4 + 2 + 2$ and 1 carried over is 9 in the tens column.

```
       (T) (U)
  ∴     4   3
        2   8
        2   4
       ─────
        9   5
        1
```

(c) $9 + 9 + 5$ is 23.

This is 2 tens and 3 in the units column.

$4 + 2 + 1$ and 2 carried over is 9 in the tens column.

$$\therefore \quad \begin{array}{cc} (T) & (U) \\ 4 & 9 \\ 2 & 9 \\ 1 & 5 \\ \hline 9 & 3 \\ \hline 2 & \end{array}$$

(d) $3 + 9 + 4$ is 16.

This is 1 ten and 6 in the units column.

$6 + 3 + 4$ and 1 carried over is 14.

This is 1 hundred, and 4 in the tens column.

1 hundred is carried over to the hundreds column.

$$\therefore \quad \begin{array}{ccc} (H) & (T) & (U) \\ & 6 & 3 \\ & 3 & 9 \\ & 4 & 4 \\ \hline 1 & 4 & 6 \\ \hline 1 & 1 & \end{array}$$

(e) $8 + 6 + 2$ is 16.

This is 1 ten and 6 in the units column.

$2 + 3 + 9$ and 1 carried over is 15.

This is 1 hundred, and 5 in the tens column.

$1 + 1 + 3$ and 1 carried over is 6.

This is 6 in the hundreds column.

$$\therefore \quad \begin{array}{ccc} (H) & (T) & (U) \\ 1 & 2 & 8 \\ 1 & 3 & 6 \\ 3 & 9 & 2 \\ \hline 6 & 5 & 6 \\ \hline & 1 & 1 \end{array}$$

12 Basic Arithmetic Operations

We are more interested in real life examples from business but the method is the same. However, we have now to decide what we can add together.

Worked example

As a freelance painter and decorator you make the following charges: £150 for decorating a room, £23 for wallpaper and £18 for paint. What is your total charge?

Answer

All the charges are in £s. Therefore they can be collected (added) together.

```
          (H)  (T)  (U)
    ∴      1    5    0
                2    3
                1    8
          ─────────────
           1    9    1
                1
```

Your total charge will be £191.

Examples 2.1

1. Add the following:

   (a) 32   (b) 62   (c) 121   (d) 231   (e) 488
       64   49   234   165   279
       11   34   179   384

2. The cost of insulating a small semi-detached house is: cavity wall insulation £215, loft insulation £145, external wall insulation £2025 and double glazing thoughout the house £1460. What is the total cost of insulation?

3. The costs of running a small business are: rates £1835, salaries £68 180, national insurance contributions £4099, heating, lighting and power £4380. How much must the company make to earn a profit of £31 000?

4. A small private hotel employs the following staff:
   2 waitresses
     one is on duty for breakfast only – 21 hours per week
     one is on duty for lunch and dinner – 42 hours per week
   1 chef who is on duty 63 hours per week
   2 chambermaids who are both on duty for 35 hours per week.
   What is the total number of hours worked by the staff?

5. The costs of making one tonne of a chemical product are:

Operating costs of producing the product	£369
Materials	£372
Fixed costs (including overheads)	£69
Plant depreciation	£58

   What is the total cost of producing one tonne of product?

2.2 Subtraction

The method of taking things away from things of the same kind is called subtraction. For example if you earn £166 per week and your employer takes away £36 for income tax and national insurance your take-home pay will be £166 less £36. This can be worked out using subtraction or 'take away'. The sign for take away is '−' and is called 'minus'. We can write the subtraction:

```
(H)  (T)  (U)
 1    6    6
      3    6
 ─────────────
```

Take 6 away from 6, that is 6−6 is 0 in the units column.

Take 3 away from 6, that is 6−3 is 3 in the tens column.

Take 0 away from 1, that is 1−0 is 1 in the hundreds column.

```
(H)  (T)  (U)
 1    6    6
      3    6
 ─────────────
 1    3    0
 ─────────────
```

14 Basic Arithmetic Operations

Worked examples

Work out the subtractions:

(a) 84 (b) 82 (c) 390 (d) 168 (e) 785
    32   69   273   72   592
    ──   ──   ───   ──   ───

    ──   ──   ───   ──   ───

Answers

(a)
$$\begin{array}{r} 84 \\ 32 \\ \hline \end{array}$$

Take 2 away from 4, that is $4-2$ is 2 in the units column.

Take 3 away from 8, $8-3$ is 5 in the tens column.

$$\begin{array}{cc} (T) & (U) \\ 8 & 4 \\ 3 & 2 \\ \hline 5 & 2 \end{array}$$

(b)
$$\begin{array}{r} 82 \\ 69 \\ \hline \end{array}$$

You cannot take 9 away from 2 because it is more than 2.

$2-9$ in the units column won't go.

If you add 10 to the 2 you get 12.

$\therefore$ $12-9$ in the units column gives 3.

But you added 1 ten to the 2 on the top therefore you must take away 1 ten in the tens column.

$8-6-1$ is $2-1$ in the tens column.

$$\begin{array}{cc} (T) & (U) \\ 8 & ^{1}2 \\ 6 & 9 \\ \hline 1 & 3 \\ \scriptscriptstyle 1 & \end{array}$$

(c)   3   9   0
            2   7   3

0 − 3 won't go; add 1 ten to 0.

∴ 10 added to 0 is 10. 10 − 3 is 7 in the units column.

9 − 7 − 1 is 2 − 1 is 1 in the tens column.

3 − 2 is 1 in the hundreds column.

	(H)	(T)	(U)
	3	9	¹0
	2	7	3
	1	1	7
		1	

(d)   1   6   8
             7   2

8 − 2 is 6 in the units column.

6 − 7 won't go; add 1 hundred to 60.

∴ 16 − 7 is 9 in the tens column.

1 − 0 − 1 is 1 − 1 is 0 in the hundreds column.

	(H)	(T)	(U)
	1	¹6	8
		7	2
	0	9	6
	1		

(e)   7   8   5
            5   9   2

5 − 2 is 3 in the units column.

8 − 9 won't go; add 1 hundred to 80.

∴ 18 − 9 is 9 in the tens column.

Basic Arithmetic Operations

$7-5-1$ is $2-1$ is 1 in the hundreds column.

	(H)	(T)	(U)
	7	¹8	5
	5	9	2
	1	9	3
	1		

Worked example

You are selling a three-piece suite marked down from £1280 to £1024. Your break-even point is £768. How much profit would you make before and after discount?

Answer

Profit before discount is £1280 − £768.

```
    1280
     768
    ____

    ____
```

$0-8$ won't go;
add 1 ten to 0 $10-8$ is 2   put 2 in units column.

$8-6-1$ $2-1$ is 1   put 1 in tens column.

$2-7$ won't go;
add 1 thousand to 200 $12-7$ gives 5   put 5 in hundreds column.

$1-0-1$ $1-1$ is 0.

```
    1280
     768
    ____
     512
     1 1
```

Profit after discount is £1024 − £768.

```
    1024
     768
    ____

    ____
```

Basic Arithmetic Operations

4 − 8 won't go;
add 1 ten to 4   14 − 8 is 6   put 6 in units column.

2 − 6 − 1 won't go;
add 1 hundred to 2   12 − 7 is 5   put 5 in tens column.

0 − 7 − 1 won't go;
add 1 thousand to 0   10 − 8 is 2   put 2 in hundreds column.

1 − 1 is 0.

$$\begin{array}{r} 1024 \\ 768 \\ \hline 256 \\ \hline {\scriptsize 1\ 1\ 1} \end{array}$$

Therefore profit before discount is £512

and profit after discount is £256.

Examples 2.2

1. Work out the subtractions:

   (a) 56   (b) 91   (c) 421   (d) 999   (e) 674
        42   65   218   981   488
        ──   ──   ───   ───   ───

        ──   ──   ───   ───   ───

2. The money taken in by a firm per week is £82 650. The expenses and salaries are £74 982 per week. What is the profit for each week before tax?

3. A garage owner has sold a car for £7800 and has allowed £1650 in part exchange for the customer's old car. How much money did he get in cash terms on the transaction?

4. Typical heating costs for heating a semi-detached bungalow using modern storage heaters are given below.
           Well insulated: £205 per annum
           Poorly insulated: £352 per annum
   What saving would be made by insulating the house properly?

5. Max Roberts earns £8562 per year. He is allowed to earn the first £3655 free of tax. On how much of his salary does he pay tax?

2.3 Multiplication

Multiplication means adding several times. For instance if you multiply 185 by 7 you add 185 seven times:

$$185 + 185 + 185 + 185 + 185 + 185 + 185$$

This gives 1295.

The sign for multiply is '×' which means 'times':

$$7 \times 185 \text{ is } 7 \text{ times } 185$$

In order to be able to do the above multiplication you need to know your multiplication tables. When you become competent in understanding what is happening the calculator can do the drudgery of this memory work. We can multiply 185×7 in the following way:

(Th)	(H)	(T)	(U)
	1	8	5
			7

Step 1 Multiply the unit number, 5, by 7 which gives 35.
Place the 5 in the units column and carry over the 3 into the tens column.

Step 2 Multiply the 8 in the tens column by 7 which gives 56. Add the 3 carried over. This gives 59.
Place the 9 in the tens column and carry over the 5 into the hundreds column.

Step 3 Multiply the 1 in the hundreds column by 7 which gives 7. Add the 5 carried over. This gives 12. Place the 2 in the hundreds column and the 1 in the thousands column.

(Th)	(H)	(T)	(U)
	1	8	5
			7
1	2	9	5
		5	3

Sometimes you have to multiply by numbers bigger than 9:

```
    321
     32
   ────
    642 ← Multiply 321 by 2.                                  (1)
  9 630 ← Put 0 in units column.
          Multiply 321 by 3 starting in tens column.          (2)
  ─────
 10 272 ← Add (1) and (2).
  ─────
    1 1
```

You can do it in the reverse order:

```
    321
     32
   ────
  9 630 ← Put 0 in units column.
          Multiply 321 by 3 starting in tens column.          (1)
    642 ← Multiply 321 by 2.                                  (2)
  ─────
 10 272 ← Add (1) and (2).
  ─────
    1 1
```

Do whichever method suits you best. This method is called *long multiplication*.

Worked examples

Do the following multiplications:

(a) 692   (b) 453   (c) 345
       8   72   124
     ───   ───   ───

     ───   ───   ───

Answers

(a)   692
         8
     ─────
     5 536   Multiply 692 by 8.
     ─────
       7 1

(b)   453
          72
       ─────
         906 ← Multiply 453 by 2.   (1)
      31 710 ← Put 0 in units column. Multiply 453 by 7.   (2)
       ─────
      32 616 ← Add (1) and (2).
       ─────
         1

(c)   345
         124
       ─────
       1 380 ← Multiply 345 by 4.   (1)
       6 900 ← Put 0 in units column. Multiply 345 by 2.   (2)
      34 500 ← Put 0 in units and tens columns.
                Multiply 345 by 1.   (3)
       ─────
      42 780 ← Add (1), (2) and (3).
       ─────
        1 1

Worked example

An employer is faced with having to give his 356 employees a weekly rise of £6. How much will this cost him in a year?

Answer

Since there are 52 weeks in a year each worker will receive:

$$\begin{array}{r} 52 \\ 6 \\ \hline 312 \\ \hline 1 \end{array}$$

Therefore the total wage bill is:

         356
         312
       ─────
         712 ← Multiply 356 by 2.
       3 560 ← Put 0 in units column. Multiply 356 by 1.
     106 800 ← Put 0 in units and tens columns. Multiply 356 by 3.
       ─────
     111 072
       ─────
        1 2

∴   The wage increase is £111 072.

Examples 2.3

1. Carry out the following multiplications:

   (a) 287　(b) 382　(c) 483　(d) 574　(e) 4562
   　　　5　　　　21　　　　64　　　　326　　　　492

2. What is the weekly wage bill for 28 people earning £158 per week?

3. The mileage allowance for a sales representative is 28 pence per mile for the first 1000 miles in a year and 24 pence per mile for the remainder. How much should he claim for 27 362 miles?

4. The average number of customers going though a checkout at a supermarket in an hour is 15. The store is open from 9.00 a.m.–7.00 p.m. for 6 days a week. How many customers go through in a week?

5. If the average amount spent by each customer is £29 what are the weekly takings from the 9 checkout points in the store?

2.4 Division

Division is the opposite of multiplication. For example 7×4 is 28.

∴ If 28 is divided by 4 it gives 7

and if 28 is divided by 7 it gives 4.

The symbol for 'is divided by' is ÷

i.e. $28 \div 7$ gives 4 and $28 \div 4$ gives 7.

Short division

Short division is division by numbers between 1 and 12, i.e. division by numbers which you meet in the multiplication tables. Again you must know these properly to do division.

Let us divide 605 024 by 7.

$605\,024 \div 7$ is worked out as follows:

$$7\,\overline{)\,605\,024}$$

It can be seen that 7 cannot go into 6.

Therefore, take 6 in with the next figure (0) to give 60.

Since 7×8 is 56, 7 will go into 60 8 times with 4 left over.

Put 8 below the first 0.

Take the 4 in with the next figure (5) to give 45.

$$7 \,|\, 60^4 5\,024$$
$$8$$

Since 7×6 is 42, 7 will go into 45 6 times with 3 left over.

Put the 6 below the 5.

Take the 3 in with the next figure (0) to give 30.

$$7 \,|\, 60^4 5\,^3 0\,2\,4$$
$$8\ 6$$

Since 7×4 is 28, 7 will go into 30 4 times with 2 left over.

Put the 4 below the second 0.

Take the 2 in with the next figure.

$$7 \,|\, 60^4 5\,^3 0\,^2 2\,4$$
$$8\ 6\ 4$$

Since 7×3 is 21, 7 will go into 22 3 times with 1 left over.

Put the 3 below the 2.

Take the 1 in with the next figure.

$$7 \,|\, 60^4 5\,^3 0\,^2 2\,^1 4$$
$$8\ 6\ 4\ 3$$

Since 2×7 is 14, 7 will go into 14 2 times with nothing left over.

Put the 2 below the 4.

The complete division is:

$$7 \,|\, 605\,024$$
$$86\,432$$

Basic Arithmetic Operations

Long Division

Long division is division by numbers larger than 12, i.e. numbers you will not meet in the multiplication tables.

Let us divide 68 768 by 28.

We do this in the following way. As in short division, divide 28 into 6. It won't go. Join the 6 to the next figure, 8, to give 68.

28 goes into 68 2 times. $28 \times 2 = 56$. Put 2 above the 8 and 56 underneath 68 as shown below and subtract:

$$\begin{array}{r} 2 \\ 28\overline{)68\,768} \\ 56 \\ \hline -12 \end{array}$$

Bring down the 7:

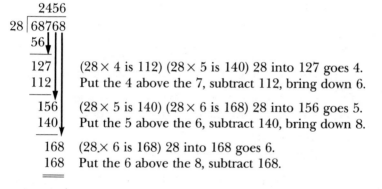

(28 × 4 is 112) (28 × 5 is 140) 28 into 127 goes 4.
Put the 4 above the 7, subtract 112, bring down 6.

(28 × 5 is 140) (28 × 6 is 168) 28 into 156 goes 5.
Put the 5 above the 6, subtract 140, bring down 8.

(28 × 6 is 168) 28 into 168 goes 6.
Put the 6 above the 8, subtract 168.

Worked examples
Do the following divisions:

(a) $2608 \div 8$ (b) $3478 \div 9$ (c) $11\,484 \div 36$ (d) $23\,957 \div 29$

Answers
(a) $2608 \div 8$

8 won't go into 2. Join 2 to the next 6 to give 26.

$$\begin{array}{r} 8\,\overline{|2608} \\ 326 \end{array}$$

8 × 3 is 24, 2 left over, put 2 with 0 to give 20.

8 × 2 is 16, 4 left over, put 4 with 8 to give 48.

8 × 6 is 48.

24 Basic Arithmetic Operations

(b) $3478 \div 9$

9 won't go into 3. Join 3 to the next 4 to give 34.

$9 \overline{\smash{\big)}\ 3478}$
        386 9×3 is 27, 7 left over, put 7 with 7 to give 77.

9×8 is 72, 5 left over, put 5 with 8 to give 58.

9×6 is 54, 4 is left over. This 4 is called a *remainder*.

(c) $11\,484 \div 36$

36 won't go into 1. Put 1 with the next 1 to give 11.

36 won't go into 11. Put 11 with the next 4 to give 114.

36 will go into 114.

```
          319
   36 )11484       (36 × 3 is 108) (36 × 4 is 144), 36 into 114 goes 3.
        108        Put the 3 above the 4, subtract, bring down 8.

         68        (36 × 1 is 36) (36 × 2 is 72), 36 into 68 goes 1.
         36        Put the 1 above the 8, subtract, bring down 4.

        324        36 × 9 is 324.
        324        Put the 9 above the 4, subtract.
```

(d) 29 won't go into 2. Put 2 with the next 3 to give 23.

29 won't go into 23. Put 23 with the next 9 to give 239.

```
          826
   29 )23957      (29 × 8 is 232) (29 × 9 is 261), 29 into 239 goes 8.
        232       Put the 8 above the 9, subtract, bring down 5.

         75       (29 × 2 is 58) (29 × 3 is 87), 29 into 75 goes 2.
         58       Put the 2 above the 5, subtract, bring down 7.

        177       (29 × 6 is 174) (29 × 7 is 203), 29 into 177 goes 6.
        174       Put the 6 above the 7.

          3       This is the remainder, subtract.
```

Examples 2.4

1. Carry out the following divisions:

   (a) $384 \div 6$ (b) $1670 \div 9$ (c) $15\,288 \div 42$ (d) $2278 \div 18$

2. A business consultant charges a daily rate for his services of £266. He works a seven-hour day. What is his hourly rate?

3. Out of his profits an employer decides to divide £8120 qually among his 56 employees as a bonus. How much does each employee receive?

4. A self-employed person pays £198 75p for class 2 National Insurance contributions in 1986–87. How much is paid per week (it is a 53 week year) in pence? (*Hint:* first change the contributions to pence.)

5. The investors in a company own 140 000 shares between them. The profits allow a dividend of £18 200 in total. How much is due on each share? (*Hint:* change £18 200 to pence.)

2.5 Combinations of the Four Rules

Before we deal with this we need to define a new term. Some calculations use *brackets* round them, for example:

$$(4+7) \times 3$$

This means that we add $4+7$ and then multiply the result by 3:

$$(4+7) \times 3 = 11 \times 3 = 33$$

The *equals* sign means that:

$$(4+7) \times 3 \text{ is the same as } 33$$

Now let us look at some examples which contain more than one rule.

$4 + 7 \times 3$	If we add first we get $4 + 7 = 11$
	If we then multiply we get $11 \times 3 = 33$
$4 + 7 \times 3$	If we multiply first we get $7 \times 3 = 21$
	If we then add we get $4 + 21 = 25$

Basic Arithmetic Operations

The order in which we carry the calculation out is important. There is a set of rules we follow called the BODMAS rule. This makes sure we get it right each time.

B   B is for brackets: if there are any figures in brackets work these out first.

O   We next multiply if 'of' appears. O is short for 'of'. You will see in the next chapter *of* means the same as multiply. It relates to fractions.

D   Divide

M   Multiply

A   Add

S   Subtract

The difficult one to understand here is 'of'. It relates to fractions.

For example   ½ of 6 is 3

This is the same as   ½ × 6 is 3

We follow the rules above but 'of' will not appear until the next chapter.

Worked example

Calculate: $6 - 8 \div 4 + 3$

Answer

$6 - 8 \div 4 + 3$ divide: $8 \div 4 = 2$

$6 - 2 + 3$ add: $6 + 3 = 9$

$9 - 2$ subtract: $= 7$

Worked example

Calculate: $(7 + 3 - 2) \times 8$

Answer

$(7 + 3 - 2) \times 8$ brackets/add: $7 + 3 = 10$

$(10 - 2) \times 8$ brackets/subtract: $10 - 2 = 8$

8×8 multiply: $= 64$

Examples 2.5

Calculate the following:

1. $7 \times 4 + 20 \div 4$

2. $(4 + 2 \times 8) \div 5$

3. $(4 - 2) \times (14 + 8) \div (5 + 6)$

4. $6 \times 7 \div 3 + 4 - 5$

5. $3 + 2 - 6 \times 12$

In business calculations you can almost always tell the order of using the rules. You will see this in Exercise 2 which covers all the rules.

2.6 Calculator Notes

You should now understand the importance of the basic rules of arithmetic. It is time to use your calculator. This makes your work easier.

Important note. Press C immediately after switching on and before any continuous sequence of calculations. If an error occurs press C and start again.

It is very simple to use. If you wish to calculate 7 + 9 press 7, then +, then 9. Finally press = for your answer (16).
For subtraction e.g. 9 − 7 press 9, then −, then 7, then = (2).
For multiplication e.g. 356 × 742 press 3, followed by 5, followed by 6.
Now press × then press 7, followed by 4, followed by 2.
Finally press = (264 152).

After pressing 356 check the screen to make sure you are correct. Always do this after putting any number into the calculator.
For division e.g. 1674 ÷ 62 press 1674, followed by ÷, followed by 62, then press = (27).
Note you can use a combination of rules provided you always use the BODMAS rule.

Basic Arithmetic Operations

Worked example

Determine $3 \times 2 + 4$ by calculator.

Answer

Press C, 3, ×, 2, +, 4, = in sequence.
$$(= 10)$$

Exercise 2

1. A car user has driven 12 132 miles for his firm. He is given a choice on which method to claim allowances.

    (a) Essential user's allowance
    Basic car allowance £795 per annum
    Up to 10 000 miles 24p per mile
    Over 10 000 miles 15p per mile

    (b) Casual user's allowance
    Up to 1500 miles 32p per mile
    Next 1500 miles 27p per mile
    Over 3000 miles 20p per mile

    Which method should he use?

2. An employer has 683 employees. He pays them £118 per week. They go on strike for a pay rise of £8 per week. The employer cannot afford to increase his salary bill. How many staff must be made redundant to keep the wages bill no higher than it is at present?
By how much would he have to increase his profit to prevent redundancies?

3. An engineering consultant charges £300 a day for his services, including expenses. He estimates his daily expenses as follows (when he is working):

    | Travelling | £47 |
    | Subsistence | £36 |
    | Overheads | £77 |

    He nominally works a seven-hour day. What does he charge per hour for his *work*?

4. The costs of running a car hire firm per year are:

Rates	£830
Electricity	£270
Telephone	£932
Car insurance	£2100
Hire purchase	£3200
Maintenance/Servicing	£1160

Altogether the cars cover 179 400 miles at 39 miles per gallon. Petrol costs 169 pence per gallon. What are the costs of running the firm?

5. The firm in question 4 brings in £1731 per week (on average). Its weekly costs are £312. Each of its 5 drivers earn £123 per week, and £11 employer's National Insurance contribution is paid per week for each driver. The firm pays the manager a weekly wage of £204 and £21 National Insurance contributions.

(a) What are the firm's total weekly costs?

(b) What is the weekly profit (on average)?

(c) What is the annual profit?

Chapter 3

Fractions

In arithmetic we do not always deal with whole numbers. We have to use fractions. Since metrication decimal fractions have become most useful but there is still a need for vulgar fractions. Both of these, and the related topic of approximations, are dealt with in this chapter.

3.1 Vulgar Fractions

Fractions are parts of a whole unit. If they are smaller than one they are called proper fractions. For example if you take a circle and divide it into four equal parts (Figure 3.1) each part is $1 \div 4$. This is written as:

$$\frac{1}{4}$$

Therefore ¼ of a circle is found by dividing the 1 by 4.

Addition and subtraction of proper fractions

Look at the rectangle in Figure 3.1. It is divided into three equal parts. Each part is $1 \div 3$. This is written as:

$$\frac{1}{3}$$

Fig. 3.1

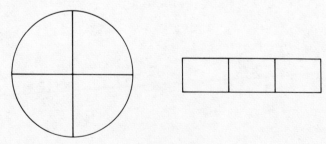

Fractions 31

Worked examples

What fraction of a circle is shown by the shaded parts and the unshaded parts in (a), (b) and (c) of Figure 3.2?

Fig. 3.2

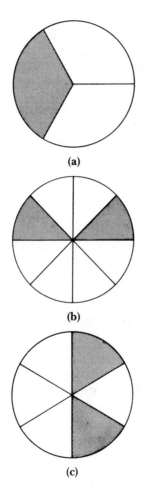

(a)

(b)

(c)

Fractions

Answers

(a) The circle is divided into three equal parts. Therefore the shaded part forms ⅓ of the circle. The unshaded part forms ⅓ + ⅓ of the circle.

$$\frac{1}{3}+\frac{1}{3}=\frac{2}{3}$$

If the bottom parts of the fractions are the same, fractions are added by adding the top parts: $1+1=2$.

The top part of the fraction is called the *numerator* and the bottom part the *denominator*.

(b) The circle is divided into eight equal parts. Two parts of this are shaded and six parts are unshaded. The shaded parts = ²⁄₈. If you look at this carefully you will see this is ¼ of the circle.

A fraction can sometimes be made simpler. You can do this by dividing the top and bottom by the same number, e.g. by 2.

In this case the shaded part $=\dfrac{\cancel{2}^{1}}{\cancel{8}_{4}}=\dfrac{1}{4}$

The unshaded part $=\dfrac{\cancel{6}^{3}}{\cancel{8}_{4}}=\dfrac{3}{4}$ (top *and* bottom divided by 2).

(c) The circle is divided into six equal parts. Two are shaded and four are unshaded:

$$\text{shaded part} = \frac{\cancel{2}^{1}}{\cancel{6}_{3}} = \frac{1}{3}$$

$$\text{unshaded part} = \frac{\cancel{4}^{2}}{\cancel{6}_{3}} = \frac{2}{3}$$

Confirm that if you add the shaded and unshaded part in each case you get an answer of 1.

It is not always easy to see whether numbers can be cancelled or not. The following rules can be useful in doing this.

Division by 10

If there are zeros on the end of a numerator and denominator they can be cancelled because you are dividing by 10, or 100, or 1000 etc.

e.g. $\dfrac{6\cancel{0}}{7\cancel{0}}=\dfrac{6}{7};\ \dfrac{75\cancel{0}}{80\cancel{0}}=\dfrac{75}{80};\ \dfrac{3\cancel{00}}{4\cancel{00}}=\dfrac{3}{4};\ \dfrac{700\cancel{0}}{850\cancel{0}}=\dfrac{70}{85}$

Fractions 33

Division by 2

If the numerator and denominator are even numbers they can be divided by 2. Sometimes they can be divided by multiples of 2 (4, 8, 16) but you may not see this:

$$\text{e.g. } \frac{\cancel{64}}{\cancel{72}} = \frac{\cancel{32}}{\cancel{36}} = \frac{\cancel{16}}{\cancel{18}} = \frac{8}{9}$$

You may have seen that $8 \times 8 = 64$ and $8 \times 9 = 72$. You could then have cancelled out by 8.

Division by 5

If the numerator and denominator end in 5 or 0 they can be divided by 5:

$$\text{e.g. } \frac{\cancel{75}}{\cancel{80}} = \frac{15}{16}; \quad \frac{\cancel{15}}{\cancel{25}} = \frac{3}{5}; \quad \frac{\cancel{75}}{\cancel{100}} = \frac{\cancel{15}}{\cancel{20}} = \frac{3}{4}$$

You may have seen in the last example that $3 \times 25 = 75$ and $4 \times 25 = 100$. You could then have cancelled by 25.

Division by 3

If the figures in a number are added together and the sum divides by 3 that number can be divided by 3:

e.g. 2304 $2 + 3 + 0 + 4 = 9$ $\therefore$ 2304 divides by 3

2379 $2 + 3 + 7 + 9 = 21$ $\therefore$ 2379 divides by 3

In a fraction both numerator and denominator must do this

$$\text{e.g. } \frac{1962}{2793} \frac{(1+9+6+2=18)}{(2+7+9+3=21)} \quad \frac{1962}{2793} = \frac{654}{931}$$

What happens if you want to add fractions with different denominators? Look at the example.

34 *Fractions*

Worked example

Add up the shaded parts in the three circles in Figure 3.3.

Fig. 3.3

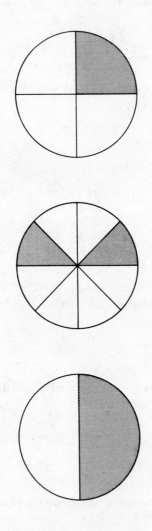

Answer

$$\frac{1}{4}+\frac{1}{2}+\frac{2}{8}$$

Fractions 35

If you cut out the shaded parts from the three circles and join them together you will form a complete circle.

$$\therefore \frac{1}{4} + \frac{1}{2} + \frac{2}{8} = 1$$

How do you carry out this addition?

Find the smallest number into which 4, 2 and 8 divide.

This is 8 since 4 × 2 gives 8
2 × 4 gives 8
8 × 1 gives 8

The number 8 cannot divide into any number less than 8. This number is called the *lowest common denominator* or the *lowest common multiple* (LCM).

Change each fraction so that it has a denominator of 8.

The denominator of ¼ is changed to 8 by multiplying 4 by 2. Since the denominator is multiplied by 2 the numerator must also be multiplied by 2.

$$\therefore \frac{1}{4} = \frac{1 \times 2}{4 \times 2} = \frac{2}{8}$$

Similarly:
$$\frac{1}{2} = \frac{1 \times 4}{2 \times 4} = \frac{4}{8}$$

$\therefore$ ²⁄₈ + ⁴⁄₈ + ²⁄₈ can now be added since they have the same denominator.

$$= \frac{2 + 4 + 2}{8}$$

$\therefore$ we can add $2 + 4 + 2 = 8$ and $\frac{8^1}{8^1} = 1$ whole circle.

Subtraction can be done in a similar way.

²⁄₃ − ⅓ is done by taking $2 - 1 = 1$ and dividing it by $3 = $ ⅓

If the denominators are different you must first obtain the lowest common denominator.

36 Fractions

Worked examples

Work out the following:

(a) $\dfrac{1}{3} - \dfrac{1}{9}$ (b) $\dfrac{1}{12} + \dfrac{2}{3} + \dfrac{1}{4}$ (c) $\dfrac{10}{18} - \dfrac{1}{2}$

Answers

(a) $\dfrac{1}{3} - \dfrac{1}{9}$

$3 \times \underline{3} = 9 \quad \therefore \dfrac{1}{3} = \dfrac{3}{9}$

$\dfrac{1}{3} - \dfrac{1}{9} = \dfrac{3}{9} - \dfrac{1}{9} = \dfrac{3-1}{9} = \dfrac{2}{9}$

(b) $\dfrac{1}{12} + \dfrac{2}{3} + \dfrac{1}{4}$

$3 \times \underline{4} = 12 \quad \therefore \dfrac{2}{3} = \dfrac{8}{12}$

$4 \times \underline{3} = 12 \quad \therefore \dfrac{1}{4} = \dfrac{3}{12}$

$\therefore \dfrac{1}{12} + \dfrac{2}{3} + \dfrac{1}{4} = \dfrac{1}{12} + \dfrac{8}{12} + \dfrac{3}{12} = \dfrac{1+8+3}{12} = \dfrac{12^1}{12^1} = 1$

(c) $\dfrac{10}{18} - \dfrac{1}{2}$

$2 \times \underline{9} = 18 \quad \therefore \dfrac{1}{2} = \dfrac{9}{18}$

$\therefore \dfrac{10}{18} - \dfrac{9}{18} = \dfrac{10-9}{18} = \dfrac{1}{18}$

Multiplication of proper fractions

Multiplication of fractions can be done as follows:

$4 \times \dfrac{2}{3}$ is the same as

$\dfrac{2}{3} + \dfrac{2}{3} + \dfrac{2}{3} + \dfrac{2}{3} = \dfrac{2+2+2+2}{3}$ which is the same as $\dfrac{4 \times 2}{3}$

Fractions

Therefore to multiply by 4 you multiply the numerator by 4.
What is a ½ of ½? It is ¼.
'Of' is another way of saying 'multiply by' or '×':

$$\frac{1}{2} \text{ of } \frac{1}{2} \text{ is } \frac{1}{2} \times \frac{1}{2} = \frac{1}{4}$$

This is obtained by multiplying numerators and multiplying denominators

$$\frac{1 \times 1}{2 \times 2} = \frac{1}{4}$$

Similarly, to find ⅔ of ⅐:

$$\frac{\text{multiply } 2 \times 1}{\text{multiply } 3 \times 7} = \frac{2}{21}$$

Worked examples

Work out the following:

(a) $\frac{1}{4} \times \frac{1}{2}$ (b) $\frac{1}{3} \times \frac{4}{5}$ (c) $\frac{3}{7} \times \frac{2}{9}$ (d) $\frac{1}{3} \times \frac{2}{5} \times \frac{2}{3}$ (e) $\frac{4}{9} \times \frac{2}{7} \times \frac{3}{4}$

Answers

(a) $$\frac{1}{4} \times \frac{1}{2} = \frac{1 \times 1}{4 \times 2} = \frac{1}{8}$$

(b) $$\frac{1}{3} \times \frac{4}{5} = \frac{1 \times 4}{3 \times 5} = \frac{4}{15}$$

(c) $$\frac{3}{7} \times \frac{2}{9} = \frac{3 \times 2}{7 \times 9} = \frac{6}{63}$$

(d) $$\frac{1}{3} \times \frac{2}{5} \times \frac{2}{3} = \frac{1 \times 2 \times 2}{3 \times 5 \times 3} = \frac{4}{45}$$

(e) $$\frac{4}{9} \times \frac{2}{7} \times \frac{3}{4} = \frac{4 \times 2 \times 3}{9 \times 7 \times 4} = \frac{24}{252}$$

Fractions

The answer to the last example can be simplified by dividing by 2 since top and bottom are even numbers:

$$\frac{24}{252} = \frac{12}{126}$$

It can be divided again by 2 since both numbers are still even:

$$\frac{12}{126} = \frac{6}{63}$$

It can be further divided by 3:

$$\frac{6}{63} = \frac{2}{21}$$

A better way is to divide (or cancel out) while the fractions are still separate. This can only be done when the fractions are joined by '×' signs.

$$\frac{\cancel{4}^1}{\cancel{9}^3} \times \frac{2}{7} \times \frac{\cancel{3}^1}{\cancel{4}^1} \qquad \begin{array}{l} \text{4 goes into 4 once} \\ \text{3 goes into 9, 3 times} \end{array}$$

$$= \frac{1 \times 2 \times 1}{3 \times 7 \times 1} = \frac{2}{21}$$

Division of proper fractions

How many quarters are there in the number 8? There are four quarters in 1 and therefore 8×4 quarters in eight.

$$\therefore 8 \div \tfrac{1}{4} \text{ is the same as } 8 \times 4$$

In other words $\div$ ¼ is the same as $\times$ ⁴⁄₁, that is ¼ turned upside down (inverted).

Therefore in division you invert the divisor (the number you are dividing by) and multiply.

For example:

$$\frac{3}{8} \div \frac{5}{8} = \frac{3}{\cancel{8}^1} \times \frac{\cancel{8}^1}{5} = \frac{3}{5}$$

$$\frac{3}{8} \div \frac{5}{8} \text{ can be written as}$$

$$\frac{\tfrac{3}{8}}{\tfrac{5}{8}}$$

If we multiply the top and bottom by the same number our fraction does not change.

Let us multiply $\dfrac{\frac{3}{8}}{\frac{5}{8}}$ by $\dfrac{8}{5}$ on top and bottom

$$\frac{\frac{3}{8} \times \frac{8}{5}}{\frac{5^1}{8^1} \times \frac{8^1}{5^1}} = \frac{3}{8^1} \times \frac{8^1}{5} = \frac{3}{5} \text{ as before.}$$

To divide, therefore, you invert the divisor and multiply.

Worked examples

Work out the following:

(a) $\dfrac{2}{3} \div \dfrac{4}{5}$ (b) $\dfrac{7}{8} \div \dfrac{3}{4}$ (c) $\dfrac{11}{45} \div \dfrac{22}{9}$

Answers

(a) $\dfrac{2}{3} \div \dfrac{4}{5} = \dfrac{2^1}{3} \times \dfrac{5}{4^2} = \dfrac{1 \times 5}{3 \times 2} = \dfrac{5}{6}$

(b) $\dfrac{7}{8} \div \dfrac{3}{4} = \dfrac{7}{8^2} \times \dfrac{4^1}{3} = \dfrac{7 \times 1}{2 \times 3} = \dfrac{7}{6}$

(c) $\dfrac{11}{45} \div \dfrac{22}{9} = \dfrac{11^1}{45^5} \times \dfrac{9^{\ 1}}{22^2} = \dfrac{1 \times 1}{5 \times 2} = \dfrac{1}{10}$

So far we have dealt only with proper fractions, i.e. those smaller than one. What about fractions bigger than one?

The answer to (b) in the worked example above is ⁷⁄₆. The numerator is greater than the denominator. This is called an *improper fraction*.

This is the same as

$$\frac{6+1}{6} = \frac{6}{6} + \frac{1}{6}$$

which is written as 1⅙ (1 and ⅙).

Fractions written as 1⅙ are called *mixed fractions*. How can we apply the four rules to these?

Addition and subtraction of mixed fractions

Consider the example:

$$1\frac{2}{3} + 2\frac{4}{5}$$

This can be rewritten as $1 + 2/3 + 2 + 4/5$.
Now add the whole numbers together $1 + 2 = 3$
and add the fractions $2/3 + 4/5$.
The lowest common denominator is 15 (3×5).

$$\frac{2}{3} \text{ is } \frac{2 \times 5}{3 \times 5} = \frac{10}{15} \text{ and } \frac{4}{5} \text{ is } \frac{4 \times 3}{5 \times 3} = \frac{12}{15}$$

$$\frac{10 + 12}{15} = \frac{22}{15} = \frac{15}{15} + \frac{7}{15} = 1 + \frac{7}{15}$$

$$\therefore 1\frac{2}{3} + 2\frac{4}{5} = 3 + 1 + \frac{7}{15} = 4\frac{7}{15}$$

Subtractions are carried out in the same way.

Worked examples

Work out the following:

(a) $3\frac{1}{4} + 4\frac{1}{6} + 2\frac{1}{12}$ (b) $8\frac{5}{9} - 4\frac{1}{3}$ (c) $3\frac{1}{6} - 2\frac{1}{3}$

Answers

(a) $3\frac{1}{4} + 4\frac{1}{6} + 2\frac{1}{12} = (3 + 4 + 2) + (\frac{1}{4} + \frac{1}{6} + \frac{1}{12})$

$$= 9 + \frac{3}{12} + \frac{2}{12} + \frac{1}{12} \qquad \text{(lowest common denominator is 12)}$$

$$= 9 + \frac{3 + 2 + 1}{12}$$

$$= 9 + \frac{6^1}{12^2}$$

$$= 9\frac{1}{2}$$

(b) $\quad 8\dfrac{5}{9} - 4\dfrac{1}{3} = (8-4) + (\dfrac{5}{9} - \dfrac{1}{3})$

$\qquad\qquad = 4 + (\dfrac{5}{9} - \dfrac{3}{9})$ (lowest common denominator is 9)

$\qquad\qquad = 4 + \dfrac{5-3}{9}$

$\qquad\qquad = 4 + \dfrac{2}{9}$

$\qquad\qquad = 4\dfrac{2}{9}$

(c) $\quad 3\dfrac{1}{6} - 2\dfrac{1}{3} = (3-2) + (\dfrac{1}{6} - \dfrac{1}{3})$

$\qquad\qquad = 1 + \dfrac{1}{6} - \dfrac{2}{6}$ (lowest common denominator is 6)

$\qquad\qquad = 1 + \dfrac{1-2}{6}$

In $\dfrac{1-2}{6}$ 2 is greater than 1. But we have no tens to carry from. We must look at it another way.

Suppose we give someone £1. Later we take £2 from him. We now owe him £1. He is short £1, i.e. he is minus £1.

$\qquad\qquad 1-2$ can then be written as -1

$\qquad\therefore \dfrac{1-2}{6} = -\dfrac{1}{6}$

$\qquad\qquad = 1 - \dfrac{1}{6}$

$\qquad\qquad = \dfrac{6}{6} - \dfrac{1}{6}$ (lowest common denominator is 6)

$\qquad\qquad = \dfrac{6-1}{6}$

$\qquad\qquad = \dfrac{5}{6}$

Multiplication and division of mixed fractions

Consider the example:

$$3\frac{3}{8} \times 2\frac{2}{9}$$

These must first be converted to improper fractions. This can be done as follows:

$$3 + \frac{3}{8} = \frac{3 \times 8}{8} + \frac{3}{8} \qquad\qquad 2 + \frac{2}{9} = \frac{2 \times 9}{9} + \frac{2}{9}$$

$$= \frac{24}{8} + \frac{3}{8} = \frac{27}{8} \qquad\qquad = \frac{18}{9} + \frac{2}{9} = \frac{20}{9}$$

$$\therefore \frac{\cancel{27}^3}{\cancel{8}^2} \times \frac{\cancel{20}^5}{\cancel{9}^1} = \frac{3 \times 5}{2 \times 1} = \frac{15}{2}$$

Convert back to a mixed number:

$$\frac{15}{2} = \frac{14 + 1}{2} = \frac{14}{2} + \frac{1}{2} = 7\frac{1}{2}$$

Division is carried out in the same way as before (invert the divisor and multiply).

Worked examples

Work out the following:

(a) $1\frac{5}{6} \times 3\frac{2}{11}$ (b) $3\frac{1}{2} \div 4\frac{1}{7}$ (c) $2\frac{1}{2} \times 4\frac{2}{11} \times 4\frac{2}{5}$

Answers

(a)
$$1\frac{5}{6} = \frac{1 \times 6}{6} + \frac{5}{6}$$

$$= \frac{6 + 5}{6} = \frac{11}{6}$$

$$3\frac{2}{11} = \frac{3 \times 11}{11} + \frac{2}{11}$$

$$= \frac{33 + 2}{11} = \frac{35}{11}$$

$$\therefore 1\frac{5}{6} \times 3\frac{2}{11} = \frac{\cancel{11}^1}{6} \times \frac{35}{\cancel{11}^1}$$

$$= \frac{35}{6} = \frac{30}{6} + \frac{5}{6} = 5\frac{5}{6}$$

(b)
$$3\frac{1}{2} = \frac{6}{2} + \frac{1}{2} = \frac{7}{2}$$

$$4\frac{1}{7} = \frac{28}{7} + \frac{1}{7} = \frac{29}{7}$$

$$\therefore \quad 3\frac{1}{2} \div 4\frac{1}{7} = \frac{7}{2} \div \frac{29}{7}$$

$$= \frac{7}{2} \times \frac{7}{29} = \frac{7 \times 7}{2 \times 29} = \frac{49}{58}$$

(c)
$$2\frac{1}{2} = \frac{4}{2} + \frac{1}{2} = \frac{5}{2}$$

$$4\frac{2}{11} = \frac{44}{11} + \frac{2}{11} = \frac{46}{11}$$

$$4\frac{2}{5} = \frac{20}{5} + \frac{2}{5} = \frac{22}{5}$$

$$\therefore \quad 2\frac{1}{2} \times 4\frac{2}{11} \times 4\frac{2}{5} = \frac{\cancel{5}^1}{\cancel{2}^1} \times \frac{46}{\cancel{11}^1} \times \frac{\cancel{22}^{2^1}}{\cancel{5}^1}$$

$$= \frac{1 \times 46 \times 1}{1 \times 1 \times 1} = 46$$

Worked example

An assistant in a veterinary practice was paid ²⁄₇ths of the money he earned for the practice. In 1986 the amount of fees brought in was £51 800. How much did the assistant get paid?

Answer

He earned:

$$\frac{2}{7} \text{ of } £51\,800$$

$$= \frac{2}{\cancel{7}^1} \times \cancel{51\,800}^{7400}$$

$$= £14\,800$$

Fractions

Worked example

A company makes £262 800 profit to be shared between three directors. The chairman receives ¼ of the profits and the remainder is split between the other directors according to their original investment. One director invests £1 000 000 and the second £800 000. How much does each director earn?

Answer

The chairman receives ¼ of 262 800

$$\frac{262\,800}{4} = £65\,700$$

The other directors invest £1 000 000 + £800 000 = £1 800 000.

$$\text{Director A invests } \frac{1\,000\,000}{1\,800\,000} = \tfrac{5}{9} \text{ of total}$$

$$\text{Director B invests } \frac{800\,000}{1\,800\,000} = \tfrac{4}{9} \text{ of total}$$

The amount to be shared = £262 800 − £65 700

= £197 100

Director A receives = ⁵⁄₉ × £197 100 = £109 500

Director B receives = ⁴⁄₉ × £197 100 = £87 600

Check: Total = £65 700 + £109 500 + £87 600 = £262 800

Examples 3.1

1. Calculate:

   (a) $\frac{2}{3} + \frac{7}{12} + \frac{1}{24}$; (b) $\frac{3}{6} \times \frac{2}{18} \times \frac{12}{17}$; (c) $\frac{13}{16} \div \frac{3}{4}$; (d) $\frac{3}{4}$ of $\frac{4}{5} \div \frac{3}{7}$

   (*Hint*: for (d) see BODMAS.)

2. Three directors share a profit of £80 000. A gets ⁵⁄₁₆, B gets ¼, C gets ⁷⁄₁₆. How much does each director get?

3. A co-operative is set up to make and sell shirts. It is agreed that the profits will be divided a follows: ⅖ will be ploughed back into the company. Each of the 19 employees will receive ²⁄₇₀ as wages. The manager will get ⁴⁄₇₀ as wages. If the profits in one year are £174 720 how will the money be divided?

3.2 Decimal fractions

As stated earlier the decimal (or denary) system puts numbers into 'blocks' of ten:

e.g. 20 is 2×10

22 is $2 \times 10 + 2$

222 is $2 \times 100 + 2 \times 10 + 2$

When we consider fractions we consider denominators in 'blocks' of 10 as well

$$\frac{1}{10} \text{ is 1 ten}\underline{\text{th}}$$

$$\frac{1}{100} \text{ is 1 hundred}\underline{\text{th}}$$

$$\frac{1}{1000} \text{ is 1 thousand}\underline{\text{th}}$$

However we do not use vulgar fractions.

We call $\quad \dfrac{1}{10} \quad\quad 0.1$

$\dfrac{1}{100} \quad\quad 0.01$

$\dfrac{1}{1000} \quad\quad 0.001$

Therefore $\dfrac{536}{1000} = \dfrac{500 + 30 + 6}{1000} = \dfrac{50\cancel{0}}{100\cancel{0}} + \dfrac{3\cancel{0}}{100\cancel{0}} + \dfrac{6}{1000}$

$$= 0.5 + 0.03 + 0.006$$

Add these. Make sure the position of the points are in line.

We get:

```
    0 . 5
    0 . 03
    0 . 006
    ─────────
    0 . 536
```

This point is called the decimal point and 0.536 is called a decimal fraction.

Fractions

Worked examples

Write down the following as decimal fractions:

(a) $\dfrac{42}{100}$ (b) $\dfrac{758}{1000}$ (c) $\dfrac{63}{1000}$ (d) $\dfrac{5842}{10\,000}$

Answers

(a) $\dfrac{42}{100} = \dfrac{40+2}{100} = \dfrac{4\cancel{0}}{10\cancel{0}} + \dfrac{2}{100} = 0.4 + 0.02 = 0.42$

Check:
0.4
0.02
$\overline{0.42}$

(b) $\dfrac{758}{1000} = \dfrac{700+50+8}{1000} = \dfrac{7\cancel{00}}{10\cancel{00}} + \dfrac{5\cancel{0}}{100\cancel{0}} + \dfrac{8}{1000}$

$= 0.7 + 0.05 + 0.008$

$= 0.758$ (check by vertical addition)

(c) $\dfrac{63}{1000} = \dfrac{60+3}{1000} = \dfrac{6\cancel{0}}{100\cancel{0}} + \dfrac{3}{1000}$

$= 0.06 + 0.003$

$= 0.063$ (check by vertical addition)

(d) $\dfrac{5842}{10\,000} = \dfrac{5000+800+40+2}{10\,000}$

$= \dfrac{5\cancel{000}}{10\,\cancel{000}} + \dfrac{8\cancel{00}}{10\,0\cancel{00}} + \dfrac{4\cancel{0}}{10\,00\cancel{0}} + \dfrac{2}{10\,000}$

$= 0.5 + 0.08 + 0.004 + 0.0002$

$= 0.5842$ (check by vertical addition)

The beauty about using decimal fractions is that they can be added, subtracted, multiplied and divided like whole numbers. But you must know where to put the decimal point.

Addition

This is carried out in the same way as the addition of whole numbers. But the decimal point is put in the same position on every line.

Worked examples

(a) 10.682 + 132.597 + 428.3000

(b) 1862.5 + 382.47 + 3961.428.

Answers

(a)
$$\begin{array}{r} 10.682 \\ 132.597 \\ 428.3000 \\ \hline 571.5790 \\ \hline \text{1 1 1} \end{array}$$

(b)
$$\begin{array}{r} 1862.5 \\ 382.47 \\ 3961.428 \\ \hline 6206.398 \\ \hline \text{2 2 1} \end{array}$$

Subtraction

This is carried out in the same way as the subtraction of whole numbers. But put the decimal point in the same position on each line.

Worked examples

(a) 632.49 − 428.38

(b) 1872.986 − 996.992

Answers

(a)
$$\begin{array}{r} 6\ 3^12.\ 4\ 9 \\ 4\ 2\ 8.\ 3\ 8 \\ \hline 2\ 0\ 4.\ 1\ 1 \\ \hline 1 \end{array}$$

(b)
$$\begin{array}{r} 1^18^17^12.^19^18\ 6 \\ 9\ 9\ 6.\ 9\ 9\ 2 \\ \hline 8\ 7\ 5.\ 9\ 9\ 4 \\ \hline \text{1 1 1 1 1} \end{array}$$

48 *Fractions*

Multiplication

This is carried out as for whole numbers but the position of the point has to be decided.

Worked example

Carry out the following multiplication:

$$18.62 \times 7.91$$

Answer

Multiply as if there were no decimal points.

```
      1862
       791
      ────
      1862    ← 1862 × 1
     16758    ← 1862 × 9 starting in 10s column
     13034    ← 1862 × 7 starting in 100s column
   ───────
   1472842
      1 1 1
```

There are two figures after the decimal point in 18.62.
There are two figures after the decimal point in 7.91.
There must be four (2 + 2) figures after the decimal point in the answer.
The answer is 147.2842.

The general rule is to add up the number of figures after the decimal point in *both* numbers. This is the number of figures after the point in the answer.

Short division

Let us look first at the following numbers:

83.2;   if we multiply by 10 we get 832.
                    Point moves 1 place to the right.
6.78;   if we multiply by 100 we get 678.
                    Point moves 2 places to the right.
49.684; if we multiply by 1000 we get 49684.
                    Point moves 3 places to the right.

By multiplying by 10 and 100 etc. we can change the fractions into whole numbers. This is done by moving the point 1 place, 2 places, etc. to the right.

In division we always have to make the number we divide by (divisor) a whole number.

Suppose you want to do the calculation:

$$5.855 \div 0.5.$$

To make the divisor 0.5 a whole number we must multiply it by 10, which gives 5.

If we multiply the bottom by 10 we must do the same to the top. We multiply 5.855 by 10 which gives 58.55.

We then do short division as with whole numbers. But the decimal point in the answer must be below the decimal point in the number being divided.

$$5 \overline{\smash{)}58.^355}$$
$$11.\ 71$$

Note position of the points.

Worked examples

Calculate the following:

(a) $58.268 \div 0.7$ (b) $513.81 \div 0.11$ (c) $7587 \div 0.9$

Answers

(a) $0.7 \times 10 = 7$
 58.268 must be multiplied by 10. It gives 582.68.

$$7 \overline{\smash{)}58^22.^16^28}$$
$$8\ 3.\ 2\ 4$$

(b) $0.11 \times 100 = 11$
 513.81 must be multiplied by 100. It gives 51381.

$$11 \overline{\smash{)}51^73^78^11}$$
$$4\ 6\ 7\ 1$$

(c) $0.9 \times 10 = 9$
 7587 must be multiplied by 10. It gives 75870.

$$9 \overline{\smash{)}75^38^270}$$
$$8\ 4\ 30$$

Fractions

Long division

In long division the same rules hold as for short division.

Worked examples

Calculate the following:

(a) $146.176 \div 3.2$ (b) $0.25575 \div 0.75$

Answers

(a) 3.2×10 gives 32
146.176×10 gives 1461.76

```
         45.68
   32 ) 1461.76
        128           32 × 4 = 128
        ───
         181
         160          32 × 5 = 160
         ───
          217
          192         32 × 6 = 192
          ───
           256
           256        32 × 8 = 256
           ═══
```

(b) 0.75×100 gives 75
0.25575×100 gives 25.575

```
         0.341
   75 ) 25.575
        225           75 × 3 = 225
        ───
         307
         300          75 × 4 = 300
         ───
          75
          75          75 × 1 = 75
          ══
```

Fractions 51

There are two difficulties which have not been mentioned.

1. The first problem concerns decimals such as 0.075 or 0.0082.

$$0.075 = \frac{0}{10} + \frac{7}{100} + \frac{5}{1000}$$

Note: $0.75 = \frac{7}{10} + \frac{5}{100}$

These are not the same

$$0.0082 = \frac{0}{10} + \frac{0}{100} + \frac{8}{1000} + \frac{2}{10\,000}$$

Again this is not the same as 0.082 or 0.82

Zeros immediately after a decimal point are important and *must not* be ignored.

Worked examples

Calculate the following:

(a) $0.0216 \div 0.8$ (b) $0.00292 \div 0.04$

Answers

(a) 0.8×10 gives 8
    0.0216×10 gives 0.216

$$8 \overline{\smash{)}0.21^56}$$
$$0.02\ 7$$

(b) 0.04×100 gives 4
    0.00292×100 gives 0.292

$$4 \overline{\smash{)}0.29^12}$$
$$0.07\ 3$$

2. When we divided in Chapter 2 we sometimes got a remainder. We do not need to leave the remainder. Sometimes it is valuable to carry it further using decimals.

   Let us consider a simple example. We wish to divide £10 between 3 people. Each would receive £3 and £1 would be left over. We would continue by dividing the £1 equally. This is easily done using decimals. Take the 10 and add, say, 3 zeros after the point:

$$3 \overline{\smash{)}10.^10^10^10}$$
$$3.\ 3\ 3\ 3$$

You can see that if we carried on we would get 3.33333 continuously. The division would never end. Clearly we must stop sometime. We do this when it no longer makes sense to continue. For example in the division above each person receives:

$$£3 + £\frac{3}{10} + £\frac{3}{100} + £\frac{3}{1000} + \underline{}$$
$$= £3 + 30p + 3p + \ldots$$

At this stage there are no longer small enough coins to continue.

Each person receives £3 and 33p or £3.33. We have taken this answer as far as two decimal places (two positions after the decimal point). We write this £3.33 correct to 2dp. Common sense dictates how many decimal places to use in real life calculations. Taking answers to a definite number of places is called approximating. We will deal with this in the next section.

Worked example

In the privatisation of a major public concern the issue was grossly oversubscribed. A member of the public applied for £15 000 worth of shares and was given £800. What fraction is this?

Answer

£800 ÷ £15000 is the same as 0.8 ÷ 15:

```
         0.0533
    15 ) 0.8000
         75        ← 15 × 5
         ──
          50
          45       ← 15 × 3
          ──
          50
          45       ← 15 × 3
```

The fraction is 0.0533 continuously. This is written as $0.05\dot{3}$. The dot above the 3 means it is repeated. The dot is called a repeater.

Worked example

The rates on a business property are £1285. The owner opts to pay in seven monthly instalments (without interest) to save money. How much will be paid monthly?

Answer

Divide the total amount into seven equal payments:

$$7 \overline{\smash{)}12^58^25.^40^50}$$
$$\phantom{7\overline{)}}1\ 8\ 3.\ 5\ 7$$

Now multiply 183.57 by 7. It can be seen that by dividing the total amount by 7 we get the monthly payment to the nearest penny *under* what is owed.

$$\therefore\ \ 183.57$$
$$\underline{7}$$
$$\underline{1284.99}$$

The owner still owes 1p. The charges must include this.

The payments will be 6 of £183.57
and 1 of £183.58

Examples 3.2

1. Carry out the following calculations:

    (a) £157.32 + £32.61 + £59.87
    (b) £182.71 − £86.53 − £21.37
    (c) £756.51 + £31.20 − £26.80 + £427.15

2. Calculate the following:

    (a) £532.41 × 27.4
    (b) £6311.07 ÷ 9
    (c) £12 660.58 ÷ 46

3. An investor has £5600.09 invested in 2467 shares. How much is each share worth?

4. An employer buys units in a managed fund for his employees on a monthly basis as part of a pension fund. He puts in £50 per month for each employee. The number of units he gets each month per employee are:

    Jan Feb Mar Apr May June July Aug Sept Oct Nov Dec
    22.81 23.92 23.87 24.01 24.12 24.72 24.16 23.98 24.51 24.81 24.92 25.16

    What is the total number of units bought for each employee in a year? In 10 years each unit will be worth £4.62. What will be the total value of the above units? (See Chapter 5 for investments.)

5. A packet of mints should weigh 30 grams and must not weigh less than 29.9 grams or more than 30.1 grams. Six million packets a week are made. What is the lowest possible weight made per week? What is the highest possible weight made per week?

6. A retailer sells a packet of mints for 11p. The average weight of each packet he sells is 29.92 grams. What is the price per 100 grams?

3.3 Approximations

The level of approximation is important in any kind of real life calculation. For example, if a builder is putting in an estimate to the local authority in bidding for a permit to build one hundred houses, he is not going to put in a bid of say:

$$£4\ 896\ 628.65$$

He is more likely to take it to the nearest hundred or even a thousand pounds. If he takes it to the *nearest* £100 he will use either:

$$£4\ 896\ 600 \text{ or } £4\ 896\ 700$$

If the figure is less than 5 in the tens column he will correct (round) downwards. If the figure is 5 or greater than 5 he will round upwards. In £4 896 628.65, this figure is 2 so he rounds down to £4 896 600.
What happens if he takes it to the nearest thousand?
He will use either £4 896 000 or £4 897 000.
Since the value in the hundreds column is 6 he will round up to £4 897 000.
If you take a commonsense viewpoint the difference

$$(£4\ 897\ 000 - £4\ 896\ 600 = £400)$$

is very small in an order of this size.

Worked examples

Approximate the following:

(a) 63.8942 to 3 dp
(b) £1684.82 to the nearest £
(c) 65.863 metres to nearest 1 tenth of a metre
(d) 35.342 kilograms to the nearest kilogram.

Answers

(a) 63.8942: fourth decimal place is 2
∴ round down to 63.894

(b) £1684.82: first decimal place is 8
∴ round up to £1685

(c) 65.863 metres: second decimal place (nearest one hundredth) is 6
∴ round up = £65.9

(d) 35.342 kilograms: first decimal place is 3
∴ round down to 35 kilograms.

An interesting example of an odd approximation occurs in money transactions. All garages quote petrol prices to a fraction of a pence. They then charge to the nearest pence. In many garages this is always rounded down. Assurance companies often quote share prices to a fraction of a pence, e.g. £2.389.

Worked example

A taxi firm runs 5 cars. Each uses an average of 31.57 litres of petrol an hour in an 8 hour day. How much petrol is used in a 7 day week? If petrol costs 37.2 pence per litre, what is the weekly cost?

Answer

In a week each car does:

$$8 \times 7 \text{ hours} = 56 \text{ hours}$$

In a week each car uses:

$$56 \times 31.57 \text{ litres.}$$

```
    31.57
       56
   ──────
   189 42
   1578 5
   ──────
   1767.92
   ──────
     1 1
```

Five cars use 1767.92 × 5 litres:

$$
\begin{array}{r}
1767.92 \\
5 \\
\hline
8839.60 \\
\hline
3\ 3\ 3\ 4\ \ 1
\end{array}
$$

Cost is 8839.60 × 37.2p:

$$
\begin{array}{r}
8839.60 \\
37.2 \\
\hline
1767\ 920\ \ \\
61877\ 20\ \ \ \\
265188\ 0\ \ \ \ \\
\hline
328833.120 \\
\hline
1\ \ \ 1\ 2\ 2\ 1\ \ \ \ \
\end{array}
$$

The cost is 328 833 pence. Since the number after the decimal point is 1 the answer is rounded down.

Answer

$$£3288.33$$

Examples 3.3

1 Approximate the following:

   (a) £51.626 to the nearest pence
   (b) 583.61 metres to the nearest ¹⁄₁₀ of a metre
   (c) 846.326 litres to the 2nd decimal place
   (d) £8462 to the nearest pound
   (e) 64.325 to the 2nd decimal place.

2. The unit prices paid by an investor for shares over a six-month period were:

          April £94.050
          May   £95.063
          June   £96.843
          July   £96.675
          Aug   £97.204
          Sept   £97.301

What was the total paid to the nearest penny?

3. A builder is assessing the price for building a house:

>   Land costs   £18 560
>   Building costs   £27 384
>   Other costs   £10 864
>   Assessed profit   £11 362

Determine the price of the house to the nearest £100.

4. 2467 shares were bought for £5683. What was the cost of a share to the nearest 0.1 pence?

5. The central heating bill for a factory was based on 32 852.621 therms per annum at 38p per therm. What is the cost for that year? Any fraction *less than* 0.5 of a penny will be disregarded.

3.4 Calculator Notes

In this chapter you have done several examples in which vulgar fractions were used.

How can you add $7/8 + 11/12$?

There are several ways:

(a) $7 \div 8$ on your calculator $= 0.875$. Write this down.

$11 \div 12$ on your calculator $= 0.91\dot{6}$.

Now $0.91\dot{6}$ is showing on your calculator 'screen'.

Add on (+) the 0.875 you have written $= 1.7917$ (appears as 1.7916666).

(b) $7 \div 8 = 0.875$ on the calculator.

On most calculators you have a memory. It may be designated as 'M in'.

Press M in (or the designated memory button).

$11 \div 12 = 0.9166666$. Press +, MR $= 1.7916666$

MR is a memory recall. It might be designated RCL. This brings 0.875 back into the calculation.

The whole process is:
7 ÷ 8
=
M in
11 ÷ 12
=
+
MR
=

(You may have M⁺ (or M+) which can be used instead of M in.)

Now try ¾ + ⅝ − ⁷⁄₁₁:
3 ÷ 4 =
M in
5 ÷ 8 =
+
MR
=
M in
7 ÷ 11 = +/− (+/− converts ⁷⁄₁₁ to −⁷⁄₁₁)
+
MR
= 0.7386364

If you have M⁺ (add to memory) and M⁻ (take from memory) you can use:
3 ÷ 4 =
M⁺
5 ÷ 8 =
M⁺
7 ÷ 11 =
M⁻
MR
= 0.7386364

Exercise 3

1. A manager in an engineering components firm is responsible for running 15 machines. He has to devote one day in every 28 days to each machine for maintenance. What fraction of the total time is required for this purpose?

2. A factory owner sets aside ⅙ of his annual returns for salaries, ⅑ for maintenance, etc, ⅓ for running costs. The rest is profit. What fraction of his return (before tax) is profit? The total income is £966 000. How much profit is made (before tax)?

3. A company has three directors who invest £28 000, £27 000 and £23 000. The last director gets a salary of £10 000 a year. The remainder of a profit of £51 000 is divided among all the directors according to their investment. How much does each director get?

4. A builder charges £57 000 for each of 10 houses if they are reserved by 1 September 1986. When the owners move in on 1 March 1987 the houses are worth £65 000. Find out the profit on each house as a decimal fraction of £57 000. What is the total profit for the 10 houses?

5. (a) A packet of mints weighs on average 30 g and sells at 10p per packet. The cost of making them is £514.91 per tonne. What is the profit per tonne? (1 tonne = 1 000 000 g.)
   (b) What fraction of the cost price of the mints is profit?

Chapter 4

Metric Measurements

In recent years there has been a very strong movement towards a decimal system of measurement. The so-called SI (short for Système International) system of units is partly based on decimal measurement. The only common quantities (units) in the SI system which have not been decimalised are time (hours, minutes and seconds) and temperature (°C). Decimal measurement is more commonly known as metric measurement and is now widespread. However, we retain a dual system in this country. Imperial measure is still used: food is sold in kilograms and pounds; petrol is sold in litres and gallons. Monetary systems are now almost universally decimalised but each country has its own currency. Even the widely-used dollar is different in value in different countries. The Australian dollar does not have the same value as the American dollar.

4.1. Measurement of Length

The basic metric unit of length is the *metre* and all other units are found by multiplying or dividing metres by 10 or multiples of ten (100, 1000, etc).

1 kilometre = 10 hectometres = 100 decametres = 1000 metres

1 hectometre = 10 decametres = 100 metres

1 decametre = 10 metres

1 kilometre is written as 1 km

1 hectometre is written as 1 hm

1 decametre is written as 1 dam

1 metre is written as 1 m

Metric Measurements

$$1 \text{ millimetre} = \frac{1}{10} \text{ centimetre} = \frac{1}{100} \text{ decimetre} = \frac{1}{1000} \text{ metre}$$
$$\text{(mm)} \qquad \text{(cm)} \qquad \text{(dm)}$$

$$1 \text{ centimetre} = \frac{1}{10} \text{ decimetre} = \frac{1}{100} \text{ metre}$$

$$1 \text{ decimetre} = \frac{1}{10} \text{ metre}$$

Note $\frac{1}{10} = 0.1$, $\frac{1}{100} = 0.01$, $\frac{1}{1000} = 0.001$

Materials are now sold by the metre as well as by the yard and they are almost equal in length. (In fact 1 metre is 3.37 inches longer than 1 yard.)

Worked examples
(a) Carpet is sold at £23 a metre length. What is the price of 1.2 decametres?

(b) A salesman is required to travel between Exeter and Bath. His journey's length going via the motorways is 223.5 kilometres one way. By road across country it is 205.8 kilometres one way. He is given a petrol allowance of 42p per kilometre. How much does he save on a *return* journey if he goes by road?

Answers
(a) 1.2 decametres = 1.2 × 10 metres = 12 metres
Cost of 12 metres = £23 × 12 = £276

(b) Motorway journey is 223.5 × 2 km = 447 km
Allowance is 447 × 42p

```
        447
         42
        ———
        894  ← 447 × 2
       1788  ← 447 × 4 (starting in tens column)
       ————
      18774
       ——
       1 1
```

18 774p = £187.74

Road journey is 205.8 × 2 km = 411.6 km

Allowance is 411.6 × 42 (normally we would use 412)

```
    412
     42
    ───
    824  ← 412 × 2
   1648  ← 412 × 4 (starting in tens column)
   ─────
  17304
   ───
    1 1
```

17 304p = £173.04

Note: (i) Although there is a saving of about £15 on a return journey it will take longer by road usually and this factor must be taken into account. (See Chapter 10.)

(ii) Using 411.6 would have given £172.87.
The difference from using 412 miles is only 17 pence.

Examples 4.1

1. Change the following values to the units asked for:
   (a) 20 metres to centimetres
   (b) 50 hectometres to kilometres
   (c) 5000 millimetres to decametres
   (d) 380 metres to millimetres
   (e) 9846 decimetres to centimetres.

2. Evaluate the cost of 8 metres of carpet at £18.59 a metre length.

3. You are buying carpet for a suite of offices. You have two quotations for the same carpet. Which is best?
   (a) 80 metres (4 m wide) at £15.20 per metre
   300 sq. metres underlay at £2.50 per sq. metre
   Gripper rods (320 m) at 35p per metre
   Seaming at £20
   Fitting at £255.
   (b) 80 metres (4 m wide) fitted with underlay at £34 per metre.

4.2 Measurement of Capacity (Volume)

The SI unit of length is the metre. The SI unit of capacity is the *cubic metre*. This can be written as m^3. However, commercially the most common unit used as the basis of measurement of volume is the *litre* which is very roughly a quarter of a gallon (1 litre = 0.22 gallon).

1 cubic metre = 1 000 000 cubic centimetres (1 m^3 = 1 000 000 cm^3).

1 litre = 1000 cubic centimetres (1000 cm^3).

$\therefore$ 1 cubic metre = $\dfrac{1\ 000\ 000}{1000}$ litres = 1000 litres = 1 kilolitre.

The other units of volume are given below in terms of litres.

1 kilolitre = 10 hectolitres = 100 decalitres = 1000 litres
(kl)   (hl)   (dal)

1 hectolitre = 10 decalitres = 100 litres

1 decalitre = 10 litres

1 millilitre = $\dfrac{1}{10}$ centilitre = $\dfrac{1}{100}$ decilitre = $\dfrac{1}{1000}$ litre
(ml)   (cl)   (dl)

1 centilitre = $\dfrac{1}{10}$ decilitre = $\dfrac{1}{100}$ litre

1 decilitre = $\dfrac{1}{10}$ litre

Worked examples

(a) What is the cost of 25.37 litres of petrol at 34.2p per litre?

(b) Which costs less: (i) 447 km at 14.1 km per litre of petrol

or (ii) 412 km at 12.3 km per litre of petrol?

Assume petrol costs 36.3p per litre. Determine the cost in each case. Is this necessary?

Answers

(a)
```
       25.37
        34.2
       -----
        5074
       10148
        7611
       -----
     867.654
```

The answer is 868p = £8.68.

64 Metric Measurements

Note: Although the cost of petrol is 34.2p per litre, the total cost must be to the nearest pence rounded *up* in this case. However some garages always round down.

(b) (i) The number of litres is $\dfrac{447}{14.1} = 31.70$ (correct to 2 dp).

$$14.1 \times 10 = 141$$

$$\therefore 447 \times 10 = 4470$$

```
            31.702
       141 ┌─────────
           │ 4470.000
             423            ← 141 × 3 = 423
             ───
             240
             141            ← 141 × 1 = 141
             ───
             990
             987            ← 141 × 7 = 987
             ───
             300
             282            ← 141 × 2 = 282
             ───
              18
```

∴ The cost is 31.70 × 36.3 = 1151p (to the nearest pence).

```
         31.70
          36.3
        ──────
          9510
         19020
          9510
        ──────
       1150710
        1 1 1
```

= £11.51

(ii) The number of litres is $\dfrac{412}{12.3} = 33.50$ (correct to 2 dp).

∴ The cost is 33.50 × 36.3 = 1216p (to the nearest p)

= £12.16 (check the working in this answer)

The cost in (i) is cheaper than in (ii). It was not necessary to work out each cost to find which costs less. The number of litres in (i) is less.

Therefore the cost is less. Multiplying by cost per litre shows *how small* the difference is.

Note: Compare this with worked example (b) on page 61. You can see that the difference in the total cost of petrol is not significant in this example.

Examples 4.2

1. Convert the following to the units asked for:

    (a) 70 centilitres to litres

    (b) 2 litres to millilitres

    (c) 3 cubic metres to cubic centimetres

    (d) 8849 cubic decametres to cubic hectometres

    (e) 569 cubic metres to litres.

2. You are importing wines. One litre bottles cost £1.95 and 70 cl bottles cost £1.40. Which is the best buy?

3. What is the cost of 25.81 litres of petrol at 37.4 pence per litre? How many miles can your car travel, at 45 miles to the gallon, on 38.20 litres of petrol? (1 litre = 0.22 gallon.)

4. Your car travels 174 miles on 19.26 litres of petrol. How many miles can be done for 1 litre?

4.3 Measurement of Weight

The basic metric weight should be the *gram* but is in fact the *kilogram* which is very roughly 2 lbs (1 kilogram = 2.2 lb).

We get the same breakdown as before into:

milligram (mg)	decagram (dag)
centigram (cg)	hectogram (hg)
decigram (dg)	kilogram (kg)
gram (g)	

Only two of these units are relevant here, the gram and the kilogram. However there is one other unit which is important. This is the tonne (or metric ton) which is defined as 1000 kg.

Metric Measurements

Worked example

A supermarket has coffee on sale at £1.63 per 100 g and £6.03 per 400 g. Which is the best buy for the customer?

If the supermarket sells 45 100 g jars and 32 400 g jars in a day what would be the value of money taken for each? What would be the total weight sold?

Answer

£1.63 for 100 g is £1.63 × 4 for 400 g = £6.52

∴ The 400 g jar at £6.03 is the better buy.

45 jars at £1.63 cost £1.63 × 45 = £73.35 (check answer).

32 jars at £6.03 cost £6.03 × 32 = £192.96 (check answer).

Number of grams = 45 × 100 + 32 × 400

$$= 4500 + 12\,800$$
$$= 17\,300 \text{ g}$$
$$= 17.300 \text{ kg}$$

Examples 4.3

1. Convert the following to the units asked for:
   (a) 8 tonnes to kilograms
   (b) 30 grams to kilograms
   (c) 56 kilograms to grams
   (d) 7564 kilograms to tonnes
   (e) 8.56 tonnes to kilograms.

2. Apples are bought from one wholesaler at 50p per kilogram and from another at 23 pence per lb. Which is the best buy?

3. Packets of fruit pastilles weighed 55 grams. They were manufactured at a rate of 5 000 000 packets a week. How many tonnes would be manufactured in a year?

Exercise 4

1. Mints are manufactured at a rate of 3400 lbs per hour. Each pack weighs 30 g. How many packs would be made in an hour? The cost of production is £518.21 per tonne. What is the production cost per hour?

2. A firm had to deliver goods to five depots. There were two possible routes which could be followed with similar traffic patterns.

   Route 1 A $\xrightarrow{10 \text{ km}}$ B $\xrightarrow{7.8 \text{ km}}$ C $\xrightarrow{11 \text{ km}}$ D $\xrightarrow{8.9 \text{ km}}$ E $\xrightarrow{6.6 \text{ km}}$ A

   Route 2 A $\xrightarrow{6.6 \text{ km}}$ E $\xrightarrow{13.9 \text{ km}}$ C $\xrightarrow{7.8 \text{ km}}$ B $\xrightarrow{12.2 \text{ km}}$ D $\xrightarrow{12.5 \text{ km}}$ A

   On paper which is the more profitable route?

3. The delivery van used for routes 1 and 2 in the previous question can do 13 kilometres to the litre. Petrol costs 37.5p per litre. What would be the estimated cost for each route?

4. A wholesaler buys 200 bottles of wine for £390. What is the cost of each bottle? Each bottle is 70 cl. She can also buy the wine in 1 litre bottles. This must be at least as cheap as the 70 cl bottles volume for volume. What is the highest price she would be willing to pay for 200 1 litre bottles?

5. The wage of each employee in a small bakery is £92.25 per week. If each person pays £12.98 National Insurance contributions and £6.40 tax what would his or her take-home pay be:

   (a) per week?

   (b) per year?

Chapter 5

Ratio, Proportion and Percentages

This chapter is extremely important. It contains much of the arithmetic required for business calculations. To make sure you understand it properly the chapter will be developed very fully and slowly.

5.1 Ratios

Let us look at Figure 5.1. This contains parts of a circle. Figure (a) is a semicircle and (b) is a circle. They both have the same radius. (The radius is the distance from the centre to the edge of the circle.)

Fig. 5.1

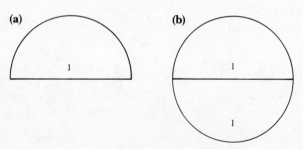

There are two semicircles in (b). Therefore (a) will go into (b) twice. (a) and (b) are said to be in a ratio of 1 to 2. Or the ratio of (a) to (b) is 1 to 2. This is written as

(a) : (b) is 1 : 2 or sometimes

(a) : (b) :: 1 : 2

Look at the figures again. (a) is ½ of (b). The ratio is only another way of writing down a fraction. Therefore 1 : 2 is the same as ½.

Worked example

A worker earns £11 675 in a year. The first £2335 is free of tax. What fraction of his salary is taxed? Express this as a ratio.

Answer

The taxed salary is £11 675 − £2335

$$= £9340$$

$$\text{The fraction} = \frac{9340}{11\,675} = \frac{4}{5}$$

The ratio of taxed salary to pretax salary is 4 to 5 *or* taxed salary : pretax salary :: 4 : 5.

Examples 5.1

1. A travelling salesman makes sales worth £80 000. His commission is £15 000. What is the ratio of sales to commission?
2. A single man earns £22 000 in a year and pays £6000 tax. What fraction of his salary does he pay in tax? Express this as a ratio.
3. An employee of a large chemical company earns £9900 a year and at the end of the year is given £550 in shares in a profit-sharing scheme. What is the ratio of his shares to his wages?

5.2 Proportion

What do we mean by proportion? Consider this example. The wages of the employees in an engineering firm are £123.

       1 employee earns £123
       2 employees earn $2 \times £123 = £246$
       5 employees earn $5 \times £123 = £615$

Therefore the total wage bill goes up *in proportion* to the number of people involved. The wage bill is directly proportional to the number of people involved.

Direct proportion

In direct proportion the problem is not always this simple. Suppose for example an employer has a total wage bill of £369 for 3 employees. He decides to increase the staff to eight all earning the same wage. What is the salary bill now? In this example there is an extra stage. We still have to find the wage of each employee.

    3 employees together earn £369

 ∴ 1 employee must earn ⅓ of £369 = £123

 ∴ 8 employees will earn 8 × £123 = £984

This method always works.

Worked example

A manufacturer making a chemical product wants to increase his capacity. The amount of product that can be made is directly proportional to the volume of the chemical plant. A pilot plant of volume 27 m^3 produces 5 tonnes of product per year. What volume is required to produce 3500 tonnes per year?

Answer

   5 tonnes are prepared in 27 m^3

 ∴ 1 tonne is prepared in ⅕ × 27 m^3 = 5.4 m^3

 ∴ 3500 tonnes are prepared in 3500 × 5.4 m^3 = 18 900 m^3

Worked example

A carpet manufacturer allowed for wastage of 1.5 m in every 100 m length of carpet. What would the wastage be for 3250 m of carpet?

Answer

The wastage is 1.5 m in *every* 100 m length. Therefore the wastage is *directly proportional* to the length of manufactured carpet.

  In 100 m the wastage is 1.5 m.

  In 1 m the wastage is $\frac{1}{100}$ × 1.5 m = 0.015 m.

  In 3250 m the wastage is 3250 × 0.015 = 48.75 m.

Ratio, Proportion and Percentages

Inverse proportion

Sometimes the kind of proportionality is different. For example an employer wants a job done in *less* time than present. She has to use *more* employees. *More* people do the same work in *less* time. As the number of people goes up, time goes down in proportion. This is called *inverse proportion*.

Worked example

A shirt manufacturer employs 35 machinists. He is given an order which would normally take 30 days. The buyer needs the order completed in 21 days. How many extra machinists does he need to bring in to achieve this?

Answer

The order takes 30 days if 35 machinists are used.

The order takes 1 day if 35×30 machinists are used $= 1050$.

The order takes 21 days if $\frac{1}{21} \times 1050$ machinists are used.

$\qquad\qquad\qquad = 50$ machinists are used.

$\therefore\ 50 - 35 = 15$ extra machinists are needed to complete the order on time.

Mixtures

Blending is a common procedure in the food, drinks and tobacco industries.

Worked example

Two coffees are ground together to form a blend. The beans of one coffee cost £1.80 per 100 g. The other beans cost £1.50 per 100 g. In what proportion would they be mixed to give a blend costing £1.60 per 100 g?

Answer

This question is normally done using algebra but it can be done by proportion, followed by trial and error as follows:

(1) £1.80 is the cost of 100 g

£1 is the cost of $\frac{1}{1.80} \times 100$ g = 55.56 g

£1.60 is the cost of $1.60 \times 55.56 = 88.90$ g.

(2) £1.50 is the cost of 100 g

£1 is the cost of $\frac{1}{1.50} \times 100$ g = 66.67 g

£1.60 is the cost of $1.60 \times 66.67 = 106.67$ g.

We can see that 88.90 g is *roughly* 12 g less than 100 g

and that 106.67 g is *roughly* 6 g more than 100 g.

∴ If 88.90 g of (1) is mixed with 2 batches of 106.67 g of (2) we get

88.90 g + 213.34 g = 302.24 g for £4.80.

If we divide 302.24/3 = 100.75 g for £1.60.

Therefore we mix 1 of (1) and 2 of (2) to get coffee costing £1.60 per 100 g.

Fortunately we have an easier and fool-proof way of working out these ratios. The mixture can be worked out as follows:

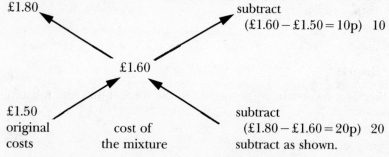

£1.80
£1.60
£1.50
original costs
cost of the mixture
subtract (£1.60 − £1.50 = 10p) 10
subtract (£1.80 − £1.60 = 20p) 20
subtract as shown.

Note: Subtract the smaller value on the left from the middle value.
Subtract the middle value from the larger value on the left.

10 to 20 is a ratio of 1 to 2.

1 part of coffee at £1.80 and 2 parts of coffee at £1.50 gives us the blend we need.

Worked example

Tobacco is mixed by blending an expensive leaf with a less expensive leaf. If the costs are £148 and £120 per kilogram how would they be blended to give a mixture costing £128 per kilogram?

Answer

Put the original prices of the tobacco leaf mixtures on the left and the price of the blend in the middle as shown:

£148

£128

£120

Subtract the smallest value from the middle value
(£128 − £120 = £8).

Subtract the middle value from the largest value
(£148 − £128 = £20).

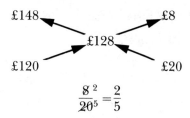

$$\frac{8}{20} = \frac{2}{5}$$

∴ 2 parts of the £148 tobacco are blended with 5 parts of the £120 tobacco.

Check:
$2 \times £148 = £296$
$5 \times £120 = £600$
───────────
$7 \times £128 = £896$

∴ The answer is correct.

Changing units and using the chain rule

It is often important to be able to change from one set of units to another. This is useful in changing from one currency to another, from litres to gallons, from feet to metres, etc.

Ratio, Proportion and Percentages

Worked example

How many dollars will you get for £750 if the exchange rate is £1 = $1.42?

Answer

$$£1 \text{ gives } \$1.42$$

$$\therefore £750 \text{ gives } \frac{1.42 \times 750}{1} = \$1065$$

Worked example

A sales representative wishes to work out his petrol costs in pence per mile. It takes 42 litres of petrol, at a cost of 37 pence per litre, to drive 350 miles.

Answer

350 miles takes 42 litres.

1 mile takes $\frac{1}{350} \times 42$ litres or $\frac{1 \times 42}{350}$ litres.

Now 1 litre costs 37 pence:

$$\frac{1 \times 42}{350} \text{ litres cost } \frac{1 \times 42}{350 \times 1} \times 37\text{p or } \frac{1 \times 42 \times 37}{350 \times 1} \text{ p}$$

This is a clumsy way to do this calculation. Let us compare it with an easier way using what we call the *chain rule* to solve this problem.

How does it work? First of all, what do we want to find?

We want to find the cost in pence. So we write this down on the left of an equals sign:

$$?\text{p} =$$

We want to find the cost in pence to drive one mile. 1 mile goes to the right of the equals sign.

$$?\text{p} = 1 \text{ mile}$$

What else is miles related to?

$$350 \text{ miles} = 42 \text{ litres}$$

What else is litres related to?

$$1 \text{ litre} = 37\text{p}$$

Ratio, Proportion and Percentages 75

We started with pence. We must finish with pence.
Put the results in a *chain*:

$$?p = 1 \text{ mile}$$
$$350 \text{ miles} = 42 \text{ litres}$$
$$1 \text{ litre} = 37p$$

The arrows link similar units. First on the right, then on the left.

Multiply the values on the right together $= 1 \times 42 \times 37$.

Multiply the values on the left together $= 350 \times 1$.

Divide the right side by the left side:

$$= \frac{1 \times 42 \times 37}{350 \times 1}$$

The complicated fraction is the same as for the long method. The final answer is 4.4p (i.e. approximately 4p per mile). By using this method we can solve this type of complicated problem every time. Let us now take a very difficult one. We can show how easily it works.

Worked example

Engineering components are manufactured at the rate of 5 200 000 per day. They sell for £16 per 1000. An export order takes 85 days to complete. What will be the cost in dollars of the order? (Exchange rate £1 = $1.28.)

Answer

What do we want:

the cost (in dollars) of the order which takes 85 days.

$$?\$ = 85 \text{ days}$$
$$1 \text{ day} = 5\,200\,000 \text{ components}$$
$$1000 \text{ components} = £16$$
$$£1 = \$1.28$$

We must end with dollars since we started with dollars:

$$\therefore \text{cost } (\$) = \frac{85 \times 5\,200\,000 \times 16 \times 1.28}{1 \times 1000 \times 1}$$

$$= \$9\,052\,160$$

This is a chain – every unit on the right must be the same as the one on the left on the next line.

Examples 5.2

1. A firm pays its manual workers £3 an hour for a 35 hour week including holidays and sick leave. If each employee has 3 weeks annual holiday, 70 hours (10 days) bank holidays and an average of 7 days per year off sick what is the actual cost to the firm for each working hour per employee?

2. An engineering components firm has agreed to carry out a rush contract in 44 working days. The task would normally take 68 days working an 8 hour day. How much overtime must each employee work to meet the deadline?

3. A secretary is required to type out a report of 48 pages. It normally takes her 15 minutes to do a page. If she gets paid £2.50 an hour, how much will she earn on the report?

4. A business consultant charges £280 a day but only expects to work 3 days a week. How much would this give her on an actual daily rate assuming a 5 day working week?

5. Tea is to be mixed from some at 80p per 100 g and some at 55p per 100 g to give a blend selling at 65p per 100 g. What is the proportion of the cheaper tea to the more expensive one?

6. Three partners invest £20 000, £35 000 and £85 000 in a business and share the profits in proportion. If the profit is £42 000 how much does each partner earn? (Hint: work out each investment as a proportion of the total amount invested.)

7. A firm proposes introducing a productivity scheme. The aim is to increase the number of components per man from 180 per day to 220 per day. How many men would be made redundant if the factory output remained the same? The daily output from the factory is 19 800 components.

Ratio, Proportion and Percentages

8. First class stamps have risen to 18p from 17p. An advertising agency sends out on average 9856 letters per week. What will be the additional cost? What would be the cost of second class post at 13p a letter? How much would be saved by using it instead of first class?

9. A salesman travels 350 miles in a day. His car travels an average of 31 miles per gallon of petrol. He pays 37.4 pence for a litre of petrol. What does petrol cost him for that day? (1 litre = 0.22 gallons.)

5.3 Percentages

Percentage values

A percentage means 'out of a 100'.

For example ½ is a fraction and means 1 part out of 2.

If instead of ½ we put $^{50}/_{100}$ this means 50 parts out of 100. This can be written as 50%.

Fig. 5.2

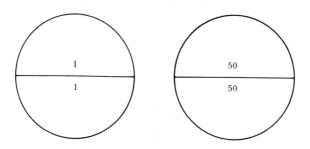

If you look at Figure 5.2 you can see that ½ = $^{50}/_{100}$

Therefore ½, $^{50}/_{100}$ or 50% are the same. It is convenient at times to express numbers as a fraction of 100 i.e. as a percentage.

It is easy to convert a number to a % by multiplying the fraction by 100.

$$\tfrac{1}{2} \times 100 = 50$$

Ratio, Proportion and Percentages

Worked examples

Express the following as percentages:

(1) Income tax at 29p in the £
(2) House rate of 180p in the £
(3) Earnings of £14 700 as a percentage of a company's profit of £140 000.

Answers

(1) The fraction of every £1 earned paid in tax

$$= \frac{29p}{£1} = \frac{29}{100}$$

$$= \frac{29}{100} \times 100\% = 29\%$$

In fractions of this kind units must be the same for comparison (in this case they are pence).

(2) The rate as a fraction is $\frac{180p}{100p}$

$$= \frac{180}{100} \times 100\% = 180\%$$

(3) Earnings as a fraction of profit are $\frac{£14700}{£140000}$

$$= \frac{147}{1400} \times 100\%$$

$$= \frac{147}{14} = \frac{21}{2} = 10\tfrac{1}{2}\%$$

Worked example

Your heating bill for an office complex is £1250 per annum. If you move from Exeter to Inverness you could expect your bill to go from 82% of the normally accepted bill to 121%. How much more would you pay?

Answer

82% is equal to £1250

∴ 1% is equal to $\tfrac{1}{82} \times$ £1250 = £15.24

121% is equal to 121 × £15.24 = £1844

The extra cost of heating = £1844 − £1250 = £594.

Ratio, Proportion and Percentages

Worked example

An employer is forced to give a 6½% pay rise to all his 60 employees. He can only afford to pay 2% if he is to retain his present profit level. How many employees would have to be made redundant for him not to go below his present profit level?

Answer

Since we are dealing in percentages we can assume each employee earns £100.

$\therefore$ The new salary = £100 + 6½% = £106.50

The owner can only pay £100 + 2% = £102.00

He can only afford £102 × 60 = £6120

Therefore if he pays £106.50 he can only afford

$$\frac{£6120}{£106.50} = 57.5 \text{ men}$$

If he employs 57 men he makes a slight gain.

If he employs 58 men he makes a loss.

$\therefore$ He must make 60 − 57 = 3 men redundant.

Worked example

Investment in managed funds is common nowadays, see Section 5.4. Units are bought at an offer price and sold at a bid price. The bid price is 5% below the offer price. This difference is to pay for management of the fund. The units therefore need to make 5% profit to recover the initial amount.

A businessman invests £85 000 in units at an offer price of £3.893. In 5 months the offer price is £4.126 per share. How many units does he buy? Fractions of units can be allocated. How much profit does he make on selling his units? Express this as a percentage of £85 000.

Answer

$$\text{The number of units} = \frac{\text{investment}}{\text{cost of each unit}} = \frac{£85\,000}{£3.893}$$

$$= 21\,834$$

When the offer price is £4.126 he can only sell at 95% of this value

$$= \frac{95}{100} \times £4.126$$

$$= £3.920 \text{ (to third decimal place)}$$

The price he gets is £3.920 for each unit.

Therefore he gets a total of £3.920 × 21 834

$$= £85\,589.28$$

∴ His profit is £85 589.28 − £84 999.76 = £589.52

This is $\frac{£589.52}{£85\,000} \times 100$ of his investment

$$= 0.69\% \text{ of } £85\,000$$

Useful ratios

There are many ratios which are useful to assess the progress of your business. These are frequently written as percentages.

(i) Performance ratios

These act as a measure of how well your business is doing. For example you can measure the efficiency of your organisation using:

$$\frac{\text{Gross profit}}{\text{Sales}}, \quad \frac{\text{Net profit}}{\text{Sales}}, \quad \frac{\text{Overheads}}{\text{Sales}}$$

where Net profit = Gross profit − Overheads

Worked example

The gross profit for a company A was £12 000 for sales of £64 000. If the overheads are £2400 determine the three ratios given above as percentages. Which is the most important as a measure of success?

Answer

$$\frac{\text{Gross profit}}{\text{Sales}} = \frac{£12\,000}{£64\,000} \times 100 = 18.75\%$$

$$\text{Net profit} = £12\,000 - £2400 = £9600$$

$$\frac{\text{Net profit}}{\text{Sales}} = \frac{£9600}{£64\,000} \times 100 = 15\%$$

$$\frac{\text{Overheads}}{\text{Sales}} = \frac{£2400}{£64\,000} \times 100 = 3.75\%$$

Net profit is the most significant value as it is determined from the other two. It is a measure of the real profit which has been made. Therefore (net profit/sales) is the most important ratio.

The rate of turnover of stock can be a measure of success. The more often you turn over your stock the greater is your sales potential. The ratio used here is:

$$\text{Stock turnover} = \frac{\text{Value of sales per year}}{\text{Value of stock}}$$

Worked example

Determine the stock turnover of a company holding £5500 of stock if the annual sale is £66 000.

Answer

$$\text{Stock turnover} = \frac{£66\,000}{£5500} = 12 \text{ times per year}$$

If the value is too low it may be that your product is too expensive or obsolete.

The easiest indicator of success to see is the ratio, return on capital.

$$\text{Return on capital} = \frac{\text{Profit before taxation}}{\text{Net assets}} \times 100\%$$

Net assets = value of company assets − value of company liabilities.

Worked example

Determine the return on capital for a company whose assets are £49 000 and liabilities are £17 000 for a year in which the profits were £4000 before tax.

Answer

$$\text{Return on capital} = \frac{£4000}{£49\,000 - £17\,000} \times 100\%$$

$$= \frac{£4000}{£32\,000} \times 100 = 12.5\%$$

(ii) Liquidity ratios

Typical ratios are:

$$\frac{\text{Current assets}}{\text{Current liabilities}}$$

$$\frac{\text{Current assets} - (\text{stock and work in progress})}{\text{Current liabilities}}$$

The second is a better test of liquidity. It only includes assets which can easily be turned into hard cash.

Worked example

A small chemical company has current assets of £15 400, stock of £2000 and work in progress to the value of £3000. Its current liabilities are £11 000. Determine the values of the liquidity ratios given above.

Answer

$$\frac{\text{Current assets}}{\text{Current liabilities}} = \frac{£15\,400}{£11\,000} = \frac{1.4}{1}$$

$$\frac{\text{Current assets} - (\text{stock and work in progress})}{\text{Current liabilities}}$$

$$= \frac{£15\,400 - (2000 + 3000)}{£11\,000} = \frac{£10\,400}{£11\,000} = \frac{10.4}{11} = \frac{0.95}{1} \text{ (to 2 dp)}$$

Other liquidity ratios are based on credit, e.g.:

$$\frac{\text{Money owed by debtors}}{\text{Annual sales}} \times 365 \text{ days } or$$

$$\frac{\text{Money owed to trade (creditors)}}{\text{Annual purchases}} \times 365 \text{ days}$$

These are measured in terms of days and obviously if too much is owed to you or by you something needs to be done. Trends are more important than single figures.

Worked example

The following figures pertain to a small bakery:

$$\text{money owed by debtors} = £550$$
$$\text{annual sales} = £5000$$
$$\text{money owed to creditors} = £600$$
$$\text{annual purchases} = £3500$$

Determine the credit/debt ratios given above.

Answer

$$\frac{\text{Money owed to creditors}}{\text{Annual purchases}} \times 365$$

$$= \frac{£600 \times 365}{£3500} \text{ days} = 63 \text{ days (to nearest day)}$$

$$\frac{\text{Money owed by debtors}}{\text{Annual sales}} \times 365$$

$$= \frac{£550 \times 365}{£5000} = 40 \text{ days (to nearest day)}$$

Examples 5.3

1. A hotel owner charges £22.50 a night for bed and breakfast + 15% VAT and 10% service charge. What is the actual price paid by the visitor?

2. The outlay of a company in 1986 was £132 800 and the return was £149 630. What was the percentage *profit*?

3. A self-employed businessman earns £28 000 in a year and has costs of £6020. Determine his profit for the year as a percentage of his earnings.

4. The profits of a small company before tax are £28 500 and tax paid is £8100. Determine the profit after tax as a percentage of the original profit.

5. A businesswoman invests £20 000 of her profit in managed funds at £1.875 per unit. How many units would she have? Later she needs capital in a hurry and withdraws her money. The *offer* price is £2.384 per unit. How much profit would she make?

84 *Ratio, Proportion and Percentages*

6. A speciality food firm makes gross profits (before tax) of £123 000 for sales of £264 000. The unsold stock is worth £8600 and the overheads are £32 000. Using performance ratios assess the efficiency of the firm's performance. If the owner could have invested his money at 11%, should he have been advised to do so?

7. A small publishing firm has current assets of £32 000 and liabilities of £18 460. Assess the liquidity of the firm (the ratio should be better than 1.5 to 1). Re-assess liquidity using £13 680 as value of stock and work in progress (the ratio should now be close to 1 to 1).

8. At the end of January 1987 a perfumery firm is owed debts of £3000 and annual sales (January to January) are £22 000. In July 1987 the appropriate figures are debts of £3800 and annual sales of £29 000. Is this a healthy trend? Give a reason for your answer.

9. The firm in question 8 owes creditors as follows:

    at the end of January 1987, £1900 for purchases (annual) of £4500;

    at the end of July 1987, £2300 for purchases (annual) of £6350.

    Again is this a healthy trend? Give a reason for your answer.

5.4 Interest and discounts

Nearly all businesses run on credit. Goods are sold on hire purchase, discounts are given for cash, etc. When estimating profits for, say, 10 years ahead you must assess their *present* value. The reason for this may not be obvious to you.

Suppose today you are going to invest in a machine worth £55 000. Let us look ahead 10 years. By that time the value of the machine will have gone down. Suppose in 10 years' time it is worth £25 000, which because of inflation will be worth less than it is now. You need to be able to work out how much this £25 000 is worth at present prices.

Suppose you invest your money instead of buying the machine. In 10 years' time it might be worth £120 000 say. Again £120 000 in 10 years' time will be worth less than it is now.

Logically, you must earn enough *profit* in the next 10 years' to offset the investment interest less the scrap value of the machine. All these figures need to be worked out at today's prices, before you decide to buy. Fortunately this can be done using interest values. We will be finding out how later in this section.

Interest

There are two types of interest payments, simple interest and compound interest. These are best shown by examples.

Ratio, Proportion and Percentages 85

(i) Simple interest

Worked example

£180 was invested at 7% simple interest for 4 years. What was the value of the investment at the end of 4 years?

Answer

When simple interest is paid the interest is on the initial investment for each year of the investment.
In this case:

$$\text{The interest on £180 at 7\%} = 180 \times 7/100 = £12.60$$

The interest at the end of year $1 = £12.60$

The interest at the end of year $2 = £12.60$

The interest at the end of year $3 = £12.60$

The interest at the end of year $4 = £12.60$

$\therefore$ The interest at the end of 4 years $= £12.60 \times 4$

$$= £50.40$$

$$\text{The interest} = \frac{180 \times 7 \times 4}{100} = £50.40$$

where £180 = amount invested (called the *principal*).

7 = %age rate of interest

4 = time (in years) of the investment

$\therefore$ Interest $= \dfrac{\text{Principal} \times \text{rate of interest} \times \text{time}}{100}$

This formula can be used in all cases of simple interest.

(ii) Compound interest

Worked example

£180 was invested at 7% compound interest for 4 years. What was the value of the investment at the end of 4 years?

86 Ratio, Proportion and Percentages

Answer

The answer at the end of the first year is the same as for simple interest.

$$\text{Interest} = £180 \times 7/100 = £12.60$$

Therefore the amount invested at the end of the first year is:

$$£180 + £12.60 = £192.60$$

In compound interest, the interest is paid on the amount at the beginning of each year.

In year 2   Interest = £192.60 × 7/100 = £13.48

Amount becomes = £192.60 + £13.48 = £206.08

In year 3   Interest = £206.08 × 7/100 = £14.43

Amount becomes = £206.08 + £14.43 = £220.51

In year 4   Interest = £220.51 × 7/100 = £15.44

Amount becomes = £235.95

$\therefore$ Total interest = £12.60 + £13.48 + £14.43 + £15.44

= £55.95

which is more than the amount gained by using simple interest.

However, doing it in this way is laborious even with a calculator. We can simplify the method.

Let us look at the example again.

At the end of year 1:

the amount = £180 + £180 × $7/10$

= £180 × 1 + £180 × $7/10$

= £180 (1 + $7/10$) since 180 is a common factor

At the end of year 2:

the amount = £180 (1 + $7/10$) + £180 (1 + $7/10$) × $7/10$

= £180 (1 + $7/10$) (1 + $7/10$)

= £180 (1 + $7/10$)^2

Ratio, Proportion and Percentages

At the end of year 3:

$$\text{the amount} = \left[£180\,(1+7/10)^2\right] + \left[£180\,(1+7/10)^2 \times 7/10\right]$$
$$= £180\,(1+7/10)^2\,(1+7/10)$$
$$= £180\,(1+7/10)^3$$

At the end of year 4 we can show the amount is $£180\,(1+7/100)^4$

i.e. $£\,\text{Principal}\,(1 + \dfrac{\text{Rate of interest}}{100})^{\text{No of years}}$

This formula can be applied to any example:

$$\text{Amount} = \text{Principal}\,(1 + \dfrac{\text{Interest rate}}{100})^{\text{No of years}}$$

These are quite complicated calculations. If you find them difficult learn the formula. The calculation is best done using a calculator. You will need a scientific calculator for this.

e.g. Amount = $£180\,(1+7/100)^4$

Stage 1 $1 + 7/100$ is $(7 \div 100) + 1 = 1.07$

(Press C, 7, ÷, 100, +, 1, =)

Stage 2   1.07 shows on your calculator.

Now press the x^y button, then 4 and =

1.310 796 shows up on your calculator

(x^y, 4, =)

Stage 3   × 180. This gives an answer of £235.94

(×, 180, =)

This is one penny different from your previous answer.

This formula always works. Let us try another example.

Worked example

A businesswoman borrows £65 000 at 12% interest. She has the choice of paying the compound interest in full after 5 years or simple interest at the end of each year. What is the difference in interest?

Ratio, Proportion and Percentages

Answer

By paying back interest at the end of each year, she is paying simple interest. Therefore at the end of 5 years the interest is:

$$\text{Interest} = \frac{£65\,000 \times 12 \times 5}{100} = £39\,000$$

Compound interest after 5 years is determined using the formula:

$$\text{Amount} = £65\,000\,(1 + \frac{12}{100})^5 = £114\,552$$

By calculator

The steps are:

   C, 12, ÷, 100, +, 1, = (1.12)

   x^y, 5, = (1.762 341 7)

(*Note:* You do not need to press = here. But it is useful to find the answer at every stage to check that you are working correctly.)

   1.762 341 7, ×, 65 000, = (114 552.21)

∴ Interest = £114 552 − £65 000 = £49 552 (to the nearest £).

∴ The difference between paying compound and simple interest here, is £49 552 − £39 000

$$= £10\,552$$

Worked example

The businesswoman in the previous example pays back £10 000 per annum in interest and in partly repaying the borrowed capital. What would she owe in 5 years?

Answer

This has to be worked out for each of the 5 years.

   Interest at the end of year $1 = 65\,000 \times {}^{12}/_{100} = £7800$

∴ Capital paid back = £10 000 − £7800 = £2200

∴ Amount owed = £65 000 − £2200 = £62 800

Interest at the end of year 2 = £62 800 × $^{12}/_{100}$ = £7536
Capital paid back = £10 000 − £7536 = £2464
∴ Amount owed = £62 800 − £2464 = £60 336

This is laborious. It is better done by calculator.

e.g. Stage 1 65 000 × $^{12}/_{100}$ = £7800

Stage 2 put £7800 in memory (M in)

Stage 3 £10 000 − take £7800 out of memory (MR) = £2200

Stage 4 put £2200 in memory (M in)

Stage 5 £65 000 − (MR) = £62 800

So we have:
$$65\,000, \times, 12, \div, 100, =$$
M in
$$10\,000, -, MR, =$$
M in
$$65\,000, -, MR, = (62\,800)$$
$$(62\,800) \times, 12, \div, 100, =$$
etc.

Repeat for 5 years using value of amount for beginning of that year instead of £65 000.

Results: Year 2 62 800   × 12 etc = 60 336
           Year 3 60 336   × 12 etc = 57 576.32
           Year 4 57 576.32 × 12 etc = 54 485.478
           Year 5 54 485.48 × 12 etc = 51 023.735

∴ She owes £51 023.74 at the end of 5 years.

If you have M+ and M− on your calculator you can use the following sequence:

$$65\,000, \times, 12, \div, 100, =, M-, 10\,000, M+, 65\,000, -, MR, =$$

Always wipe out the memory at the end of each calculation. This can be done by turning the calculator off.

Appreciation

Some property appreciates over a number of years and this can be given a certain % value using simple interest or compound interest. It is usually the latter.

Worked example

A shop was bought for £65 000. Its value appreciates at 14% per annum. What would its value be in 8 years?

Answer

This problem is solved using the compound interest formula.

$$\text{Amount} = £65\,000\,(1 + {}^{14}/_{100})^8$$

Using a calculator:

14, ÷, 100, +, 1, = (1.14)

(1.14) x^y, 8, = (2.852 586 4)

(2.852 586 4) ×, 65 000, = (185 418.12)

$$= £185\,420 \text{ (to the nearest £10)}$$

Depreciation

This is the opposite of appreciation. For example, furniture, machinery and transport go down in value. There are two ways of measuring depreciation:

(i) the equal instalment method

(ii) the diminishing balance method.

(i) The equal instalment method

An equal amount is written off the value of the asset every year.

Worked example

A public analyst buys a new piece of equipment for £18 000. It has an estimated lifespan of 6 years with a scrap value of £1200 at the end of it. What is the annual estimated depreciation? The depreciation occurs by the same value every year.

The depreciation over 6 years = initial cost − scrap value

$$= £18\,000 - £1200$$

$$= £16\,800$$

Ratio, Proportion and Percentages

The depreciation is the same every year.

$$\therefore \text{ Annual depreciation} = \frac{£16\,800}{6}$$

$$= £2800$$

It can be seen from the above that:

$$\text{Annual depreciation} = \frac{\text{cost price} - \text{scrap or residual value}}{\text{estimated number of years of use}}.$$

You will have seen in this method some resemblance to simple interest.

(ii) The diminishing balance method

This method is very similar to compound interest. A percentage of the value is *written off* the current value of the equipment each year. Instead of increases in value we get decreases in value. We can use the formula for compound interest with a minus instead of a plus:

i.e. $\quad$ Amount = Cost price $(1 - \frac{\text{Rate of depreciation}}{100})^{\text{Years}}$

Worked example

A chocolate manufacturer buys new multipacking line equipment for £330 000. This depreciates at 22% per annum. What will be its value in 9 years' time?

Answer

$$\text{Value in 9 years} = £330\,000\,(1 - {}^{22}/_{100})^9$$

Using a calculator:

$$22, \div, 100, =, +/-, +, 1, = (0.78)$$

The +/− changes 0.22 to −0.22.

$$(0.78)\ x^y,\ 9, = (0.106\,868\,9)$$

$$(0.106\,868\,9) \times, 330\,000, = (35\,266.744)$$

$$= £35\,270 \text{ (to nearest £10)}$$

Alternatively, you can use the program:

$$22, \div, 100, \text{M}-, 1, +, \text{MR}, =, x^y, 9, \times, 330\,000,$$

$$= £35\,270 \text{ (to nearest £10)}$$

Discounts

There are several types of discount. We shall look at three of them.

(i) Cash discount

Cash discounts are given when goods are paid for in cash. These are not as common as they once were since large discounts are given on many items, particularly at discount stores and supermarkets. Cash discounts can vary and there are no set rules.

Worked example

A large electrical combine is selling washing-machines at £340. They allow a 5% discount for cash. How much will each washing-machine sell for?

Answer

$$\text{Discount} = £340 \times {}^{5}/_{100}$$
$$= £17$$
$$\therefore \text{Customer pays } £340 - £17$$
$$= £323$$

(ii) Trade discount

These are given by wholesalers to retail stores. When wholesalers sell to the public, the public do not usually get this discount. Builders' merchants give this type of discount.

Worked example

A builders' merchant sells a £1650 set of kitchen units to a builder at 33% trade discount. What is the discount worth in cash?

Answer

$$\text{Discount} = £1650 \times {}^{33}/_{100}$$
$$= £544.50$$

(iii) Settlement discounts

These are paid normally to customers who pay their accounts (e.g. monthly accounts) on time. They are usually fairly small.

Discounted cash flow

You are going to invest in new machinery. Is it worth it? How long will it take for profits to recoup the cost? Would it be better to invest the money? How much interest could you expect to make? These are very difficult questions to answer. In assessing whether to buy or not you must:

(1) make a reasonable estimate of the likely percentage interest you will make by investing the money;

(2) try to assess the profits you will make, say, in the first 10 years.

You must also try to determine the value of profits in the future at today's prices. How can we do this?

Let us assume that you invest money at 11%. Your profit in three years' time is £1. What is its value *now*?

We can work this out by using the compound interest formula.

$$£1 = \text{Principal} \left(1 + \frac{11}{100}\right)^3$$

where principal is today's value.

$$\therefore \text{Principal} = \frac{£1}{(1 + \frac{11}{100})^3}$$

Using a calculator:

$$11, \div, 100, +, 1, = (1.11)$$

$$(1.11) \; x^y, 3, = (1.367\,631)$$

$$1/x, = (0.731\,191\,3)$$

$$= £0.731\,2$$

£1 in 3 years' time is worth £0.731 2 *now*.

Figure 5.3 on page 94 gives these values for interests ranging from 1% to 25% over a period of 15 years.

Suppose you make a profit of £1500 in 3 years' time its value now

$$= £1500 \times 0.7312 = £1096.80$$

The method can be clearly shown by a worked example.

Fig. 5.3

Present value of £1

Number of Years

Interest rate %	1	2	3	4	5	6	7	8	9	10	11	12	13	14	15
1	.9901	.9803	.9706	.9610	.9515	.9420	.9327	.9235	.9143	.9053	.8963	.8874	.8787	.8700	.8613
2	.9804	.9612	.9423	.9238	.9057	.8880	.8706	.8535	.8368	.8203	.8043	.7885	.7730	.7579	.7430
3	.9709	.9426	.9151	.8885	.8626	.8375	.8131	.7894	.7664	.7441	.7224	.7014	.6810	.6611	.6419
4	.9615	.9246	.8890	.8548	.8219	.7903	.7599	.7307	.7026	.6756	.6496	.6246	.6006	.5775	.5553
5	.9524	.9070	.8638	.8227	.7835	.7462	.7107	.6768	.6446	.6139	.5847	.5568	.5303	.5051	.4810
6	.9434	.8900	.8396	.7921	.7473	.7050	.6651	.6274	.5919	.5584	.5268	.4970	.4688	.4423	.4173
7	.9346	.8734	.8163	.7629	.7130	.6663	.6227	.5823	.5439	.5083	.4751	.4440	.4150	.3878	.3624
8	.9259	.8573	.7938	.7350	.6806	.6302	.5835	.5403	.5002	.4632	.4289	.3971	.3677	.3405	.3152
9	.9174	.8417	.7722	.7084	.6499	.5963	.5470	.5019	.4604	.4224	.3875	.3555	.3262	.2992	.2745
10	.9091	.8264	.7513	.6830	.6209	.5645	.5132	.4665	.4241	.3855	.3505	.3186	.2897	.2633	.2394
11	.9009	.8116	.7312	.6587	.5935	.5346	.4817	.4339	.3909	.3522	.3173	.2858	.2575	.2320	.2090
12	.8929	.7972	.7118	.6355	.5674	.5066	.4523	.4039	.3606	.3220	.2875	.2567	.2292	.2046	.1827
13	.8850	.7831	.6931	.6133	.5428	.4803	.4251	.3762	.3329	.2946	.2607	.2307	.2042	.1807	.1599
14	.8772	.7695	.6750	.5921	.5194	.4556	.3996	.3506	.3075	.2697	.2366	.2076	.1821	.1597	.1401
15	.8695	.7561	.6575	.5718	.4971	.4323	.3759	.3269	.2843	.2472	.2149	.1869	.1625	.1413	.1229
16	.8621	.7432	.6407	.5523	.4761	.4104	.3538	.3050	.2630	.2267	.1954	.1685	.1452	.1252	.1079
17	.8547	.7305	.6244	.5337	.4561	.3898	.3332	.2848	.2434	.2080	.1778	.1520	.1299	.1110	.0949
18	.8475	.7182	.6086	.5158	.4371	.3704	.3139	.2660	.2255	.1911	.1619	.1372	.1163	.0985	.0835
19	.8403	.7062	.5934	.4987	.4190	.3521	.2959	.2457	.2090	.1756	.1476	.1240	.1040	.0876	.0736
20	.8333	.6944	.5787	.4823	.4019	.3349	.2791	.2326	.1938	.1615	.1346	.1122	.0935	.0779	.0649
21	.8264	.6830	.5645	.4665	.3855	.3186	.2633	.2176	.1799	.1486	.1228	.1015	.0839	.0693	.0573
22	.8197	.6719	.5507	.4514	.3700	.3033	.2486	.2038	.1670	.1369	.1122	.0920	.0754	.0618	.0507
23	.8130	.6610	.5374	.4369	.3552	.2888	.2348	.1909	.1552	.1262	.1026	.0834	.0678	.0551	.0448
24	.8065	.6504	.5245	.4230	.3411	.2751	.2218	.1789	.1443	.1164	.0938	.0757	.0610	.0492	.0397
25	.8000	.6400	.5120	.4096	.3277	.2621	.2097	.1678	.1342	.1074	.0859	.0687	.0550	.0440	.0352

Worked example

You wish to buy a new piece of production equipment at £12 000. You envisage that a reasonable rate of interest will be paid if the money is invested at 14%. You estimate the following profits:

Year 1	Year 2	Year 3	Year 4	Year 5	Year 6
£1800	£2500	£3400	£4900	£6600	£7800

How soon will you make more money than if you had invested it at 14%?

Answer

Year	Anticipated profit		Discounted factor at 14% (See Fig. 5.3)		Discounted profit (to nearest £1)
1	£1800	×	0.8772	=	£1579
2	£2500	×	0.7695	=	£1924
3	£3400	×	0.6750	=	£2295
4	£4900	×	0.5921	=	£2901
5	£6600	×	0.5194	=	£3428
6	£7800	×	0.4556	=	£3554

The total discounted profit after 5 years = £12 127.

This is more than the £12 000 paid for the machinery.

Investment

We will only consider three forms of investment and these only briefly.

(i) Managed funds

Investors put money in a managed fund instead of buying stocks and shares. The managers use the funds on the investors' behalf by dealing on the stock market. The money is invested as units, these being priced at a level appropriate to the market. Investors are not restricted to buying whole units. Buyers buy at an 'offer' price and sell at a 'bid' price which is usually 5% lower. This 5% takes care of management's and agents' initial fees although there is usually a management charge of about ¼% per year.

96 Ratio, Proportion and Percentages

Worked example

An investor buys £60 000 worth of units at £2.569 per unit. She sells them six months later at an *offer* price of £2.976 per unit. What price does she get? What is her percentage profit?

Answer

The number of units = £60 000/£2.569 = 23 355.39

When she sells them she must sell at the *bid* price. This is 5% below the

offer price = $^{95}/_{100}$ × £2.976   = £2.827

∴ The bid value of her units = £2.827 × 23 355.39

   = £66 025.69

∴ The profit   = £6025.69

∴ % profit   = £6025.69 × 100/£60 000

   = 10.0

(ii) Shares

Shares are bought by investors usually through a brokerage firm. The firm charges a commission on each transaction. The amount of commission was, until recently, controlled but the government have arranged with the stock exchange that these controls be removed. Commission is now charged in different ways.

The value of shares may fluctuate. They can go up or down. Therefore the shareholder can make a profit or loss depending on when he or she buys and sells the shares. Dividends may also be paid at definite intervals, e.g. yearly. Therefore profits can be made in two ways. The price of a share when first issued is called the *par* value of the share. This is frequently £1.

Worked example

An investor buys oil shares (worth £1 at par) for 713.5p and sells them at 808p. Her original investment was £10 000. How much profit was made? Express this as a percentage of the original investment. Assume 0.80% of the nominal value of the share commission for the broker on each transaction.

Answer

Cost of each share	$= 713.5 + 0.80\%$ of £1 (commission)
	$= 713.5 + 0.8\text{p}$
	$= 714.3\text{p}$
$\therefore$ Number of shares	$= \dfrac{1\,000\,000\text{p}}{714.3\text{p}} = 1400$ shares
Price of shares when sold	$= 808 - 0.80\%$ of £1 (commission)
	$= 808 - 0.8\text{p}$
	$= 807.2\text{p}$ per share
$\therefore$ Total selling price	$= 1400 \times 807.2\text{p}$
	$= £11\,300.80$
$\therefore$ Profit	$= £11\,300.80 - £10\,000$
	$= £1300.80$
$\therefore$ % age profit	$= \dfrac{£1300.80}{£10\,000} \times 100$
	$= 13\%$

Worked example

An 8% dividend was paid out on the shares when held by the investor in the previous example. What was the increased percentage profit?

Answer

The nominal value of each share is £1 and the dividend is paid on this value.

$$1400 \text{ shares} = £1400$$

$$\therefore \text{ Dividend} = \dfrac{£1400 \times 8}{100}$$

$$= £112$$

$\therefore$ Total profit $= £1300.80 + £112 = £1412.80$

$$\therefore \text{ \% profit} = \dfrac{£1412.8 \times 100}{£10\,000} = 14.13\%$$

Important note: You must not add 13% + 8%. The profits are based on different values.

(iii) Stocks

Stocks are similar to shares. They are normally sold in blocks of £100 although any number can be bought. They have a fixed rate of interest and can be repayable on a fixed date. If you had, for example, Treasury 10% 1990 stocks these were loaned to the Treasury at a 10% rate of interest to be repaid in 1990. Sometimes stocks have no repayable date. However they can be bought and sold like shares. They do not have monetary values like shares but percentage values. £100 stock at 95 means the value of the stock is £95. That is, its value is 95% of the original value. Again brokerage is charged on buying and selling.

Mark up and margin

Profits are assessed in relation to cost price or selling price:

$$\% \text{ profit} = \frac{\text{selling price} - \text{cost price}}{\text{cost price}} \times 100$$

$$\text{or} \quad = \frac{\text{selling price} - \text{cost price}}{\text{selling price}} \times 100$$

Suppose a coat costs £60 to the retailer and sells at £90:

$$\% \text{ profit (on cost price)} = \frac{£90 - £60}{£60} \times 100 = \frac{30}{60} \times 100 = 50\%$$

$$\% \text{ profit (on selling price)} = \frac{£90 - £60}{£90} = \frac{30}{90} \times 100 = 33\tfrac{1}{3}\%$$

The profit is the same but the use of percentage values can be very different. The profit as a percentage of the cost price is called the *mark-up* value. The profit as a percentage of the selling price is called the *margin* of profit. When an item is reduced in price it is often *marked down* as a percentage of the selling price.

Worked example

A furniture manufacturer charges a retailer £750 for a three-piece suite. The retailer in turn sells this for £1000. What is the mark-up and margin? One of the suites is put on sale at £800. What is the percentage by which it is marked down? What is the new margin?

Answer

Profit	$= £1000 - £750 = £250$
Mark-up	$= \dfrac{£250}{£750} \times 100 = 33\frac{1}{3}\%$
Margin	$= \dfrac{£250}{£1000} \times 100 = 25\%$
'Loss'	$= £1000 - £800 = £200$
Mark-down	$= \dfrac{£200}{£1000} \times 100 = 20\%$
New margin	$= \dfrac{£800 - £750}{£800} \times 100$
	$= \dfrac{£50}{£800} \times 100$
	$= 6.25\%$

Examples 5.4

1. A company borrows £29 000 from the bank at 14% compound interest payable in full in 6 years. How much will be owed at that time?
2. A bakery employs 50 people at £83 per week. A pay rise of 3% is awarded. The bakery has to borrow enough money to pay for the rise over 2 years at 14% compound interest. How much will be borrowed?
3. A young man is setting up in business. He needs £35 500. A friend loans him the money at $8\frac{3}{4}\%$ simple interest.
   (a) What interest does he pay per year?
   (b) How much would he pay in six years?
4. A company director is trying to determine whether to expand his business or invest the money at 9% compound interest. He wishes to invest £92 000 for an expected profit over 10 years of £130 000 (assume tax on both is the same). Which is the better investment?
5. The investment in question 4 is subject to tax relief on £50 000 of the investment at 29% but not on the profit. How much better off is the director by investing his money?
6. A factory was built for £212 000 and was sold 8 years later. It had appreciated by 16% per annum compound interest. What was the final value?

7. The furniture in an office was worth £12 800. It was considered to depreciate by 10% every year for tax purposes. What value could be offset against tax for the first three years?

8. A company sells washing-machines at £328 but offers a discount of 5% for cash. What is the total discount on 24 machines sold for cash in one week?

9. A builder buys bathroom equipment valued at £8300 from a builders' merchant at 28% discount. How much will it cost him?

10. An engineering company is considering installing machinery worth £17 850. It will have to borrow money at 14%. Anticipated profits would be:

Year 1	£1400
Year 2	£2100
Year 3	£3200
Year 4	£6200
Year 5	£8300
Year 6	£9200
Year 7	£8800

    Would the company recoup its money within these 7 years?

11. A business woman invested £60 000 of a profit made in 1980 in the following way to help with her tax burden. Which was the best investment? (Ignore brokerage costs which were comparable in each case.)

    (a) £20 000 in a managed fund. Units were bought at £2.596. At the end of 5 years they were worth £4.751 (offer price).

    (b) £20 000 in shares at 382p (100p at par). In 5 years they were worth 512p. The dividends paid out were:

Year	1	2	3	4	5
Dividend	4%	—	3½%	8%	9%

    (c) £20 000 in 9¾% stock at 89½. In 5 years the stock sells at 110.

12. A wholesaler sells refrigerators to a retailer at £88.50 each. The retailer sells them at £118.75. What are the mark-up and margin values?

    Unfortunately they do not sell and he has to cut his price by £12. What is the mark-down? What are the new mark-up and margin values?

Ratio, Proportion and Percentages 101

Exercise 5

1. A company pays its 12 employees £103 per week. A pay rise of 3¼% is awarded. What would be the salary bill in the following year?

2. A company sells its products for £83 000 in a given year. The costs of the company are £38 800. In order to increase profits the company looks at two alternatives:

    (a) Should it try to increase sales by 10%?

    (b) Should it increase prices by 3%?

    Assume costs rise in proportion to sales. Work out the new profit in each case. What seems the best approach to use?

3. The 36 workers in a factory earn £120 a week for a 5 day week, and time and a half for overtime. A rush job comes in which would normally take 54 days, but it must be done in 45 days.

    (a) How much overtime must be worked?

    (b) If the normal day is 8 hours how much overtime will be earned?

4. A self-employed person has to pay 29% tax on £14 358 of earnings. How much does this amount to?

5. A prospective pub owner has two alternatives when setting up in business. She needs £104 300 for buildings, pumps, furniture and initial stock. There are two possible sources of income:

    (a) A brewery who will lend the money *without* interest if she sells only their products;
    (b) A bank who will lend her the money at 14.75%.

    Work out the interest payable per annum in (b). Can you suggest which she might choose, and why? (*Hint* What, roughly, is the weekly interest in (b))?

6. A retailer sells tape-recorders at £64. The cost to the retailer is £56. Express the profit as a percentage of the buying and of the selling price. Which is the margin? What is the other called?

7. A company has current assets of £30 000 and current liabilities of £19 800.

    (a) What are its net assets?

Its gross profits are £9800, the overheads are £6870 and annual sales are £23 800.

(b) Write down three ratios which can be measures of the company's performance. Which is the most significant?

The stock is maintained at £1831.

(c) What is the stock turnover? The previous year it had been 11 times per year. What can you deduce from these figures?

The stock and work in progress is £9870.

(d) Determine the liquidity ratios. Are the values in the right region?

8. A company has annual sales of £28 600 and debts owed to it of £7280. It makes purchases of £20 600 and owes £6300 to trade creditors. Write down the liquidity ratios which are the measure of:

(a) credit given

(b) credit taken.

These are both higher than the year before. What can you deduce from this?

9. A builder buys kitchen furniture worth £3825 from a builders' merchant at 33% discount and charges it at £4000 in the cost of a house. What is his percentage profit? State this as a mark-up value and a margin value.

10. A paper-making firm wants to install new machinery at a present cost of £134 000. The estimated interest on the machinery payable to a bank is 13%. The estimated profits are:

Year 1	£19 500
Year 2	£21 300
Year 3	£25 260
Year 4	£32 850
Year 5	£40 080
Year 6	£49 700
Year 7	£60 840
Year 8	£62 940
Year 9	£64 380
Year 10	£66 570

(a) When does the firm break even?

(b) What is the profit at the end of 7 years on today's prices?

11. A consultant invests his £30 000 profit in two ways:

(a) £15 000 in a managed fund at £1.879 per unit. In 2 years he sells at £2.562 (*bid* price). 10% of his original investment is free of tax. The remainder of the profit is subject to tax at 11%.

(b) £15 000 in 9¼% stock at 96 (profits are tax free).

What is his profit at the end of year 1?

What is his profit at the end of year 2?

(*Hint:* Only profit in stock is obtained at the end of year 1.)

Chapter 6

Areas and Volumes

It is frequently necessary in industry and commerce to use measurements of areas and volumes. The problems that face companies may be very varied. At one end of the scale it could be the cost of carpeting an office. At the other, it could be the best design for a building. It could involve a choice between one large technical unit as opposed to several small ones. It might be the best and most cost-effective way of packaging a chocolate bar. In most cases the calculations involve simple area and volume measurements like those given below.

6.1 Measurement of Area

The standard unit of area is the m^2 (square metre). This is the area of a square 1 metre long and 1 metre wide (Figure 6.1).

Fig. 6.1

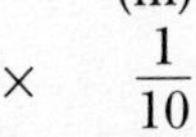

Other areas are measured in the following units

length × breadth	length × breadth (m)		area	area (m^2)
1 dm × 1 dm	$\frac{1}{10}$ ×	$\frac{1}{10}$	1 dm^2	$\frac{1}{100}$ m^2
1 cm × 1 cm	$\frac{1}{100}$ ×	$\frac{1}{100}$	1 cm^2	$\frac{1}{10\,000}$ m^2
1 mm × 1 mm	$\frac{1}{1000}$ ×	$\frac{1}{1000}$	1 mm^2	$\frac{1}{1\,000\,000}$ m^2
1 dam × 1 dam	10 ×	10	1 dam^2	100 m^2
1 hm × 1 hm	100 ×	100	1 hm^2	10 000 m^2
1 km × 1 km	1000 ×	1000	1 km^2	1 000 000 m^2

Rectangles

A rectangle 3 m × 2 m = 6 m² (see Figure 6.2(a)).

A rectangle 4 m × 5 m = 20 m² (see Figure 6.2(b)).

Fig. 6.2

(a) 3 m / 2 m

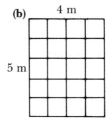
(b) 4 m / 5 m

Every rectangle has an area in which:

$$\text{area} = \text{length} \times \text{breadth}$$

Worked example

What is the area of a room 6 metres long by 10 metres broad?

Answer

Area of room = 6 m × 10 m
= 60 m²

Worked example

The room in the previous example is to be carpeted. The carpet fitter will use carpet 4 m wide to carpet the room. What is the cheapest way of fitting? How much carpet is required in each case?

Answer

There are two ways of fitting the carpet shown in Figure 6.3(i) and (ii).

Fig. 6.3

(i)

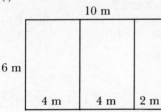

Three pieces of carpet
6 m long are required
= 3 × 6 m = 18 m
Area of carpet required
= 18 × 4 m = 72 m²

(ii)

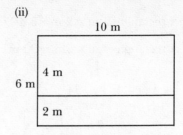

Two pieces of carpet
10 m long are required
= 2 × 10 m = 20 m
Area of carpet required
= 20 × 4 m = 80 m²

∴ (i) is the cheapest way of fitting.

Triangles
Fig. 6.4

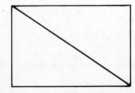

Take the rectangle in Figure 6.4 and draw a diagonal. This divides the rectangle into two equal parts.

Each part is a triangle.

The area of the triangle = ½ area of rectangle

= ½ (length × breadth)

It is easily shown from this that in the triangle below

Fig. 6.5

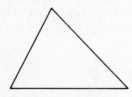

area of a triangle = ½ (length × height)

Worked example

Determine the area of a triangle of length 60 cm and height 40 cm.

Answer

Area = ½ (60 × 40) cm²

= 1200 cm²

Circles

Fig. 6.6

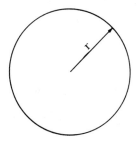

If a circle has a radius of r (see Figure 6.6)

then the area of circle = π × (radius)² = π r²

where π is a constant = 3.142 (correct to 3 decimal places).

Worked example

It is necessary to measure the area of the bottom of a cider bottle for stacking purposes. The diameter is 10 cm.

Answer

The area = π r²

= π (5)² cm² since radius = ½ diameter

= 25π cm²

= 78.6 cm²

Areas and Volumes

Examples 6.1

1. Determine the areas of the following figures:

   (a) A rectangular plot 8 m long and 3 m wide

   (b) A triangle 15 cm long and 7 cm high

   (c) A circle of radius 7 cm.

2. Each bedroom in a 12 bedroom hotel has a floor 4½ m wide and 5 m long. Work out the least expensive method of carpeting the floor using carpet 2 m wide.

3. A cylindrical tower in a chemical factory has to be restricted to a square area 35 m × 35 m. What is the largest area that the floor of the tower can cover?

6.2 Measurement of Perimeter Length

The perimeter is the length of the sides of a figure.

Rectangle

$$\text{Perimeter} = 2 \times \text{length} + 2 \times \text{breadth}$$

Let us consider a rectangle of area 16 m^2.

This could be a square = 4 m × 4 m.

It could also be several rectangles.

For example, 8 m × 2 m or 16 m × 1 m.

Measure the perimeter length in each case.

4 m × 4 m	perimeter = 2 × 4 + 2 × 4 = 16 m
8 m × 2 m	perimeter = 2 × 8 + 2 × 2 = 20 m
16 m × 1 m	perimeter = 2 × 16 + 2 × 1 = 34 m

It can be seen from this that, for the same area, a square has the shortest perimeter. This can have important financial implications, e.g. in building.

Worked example

A factory owner wants to build a store of 49 m² ground area but with the smallest building cost. This is obtained by having the smallest amount of wall space. What is the best shape for the building? What is the perimeter of the wall?

Answer

A square shape will have the smallest perimeter of wall.

An area of 49 m² can form a square 7 m by 7 m.

The perimeter $= 2 \times 7$ m $+ 2 \times 7$ m $= 28$ m

Circles

The perimeter of a circle is $\pi \times$ diameter of the circle. Since the perimeter is the circumference:

$$\text{Circumference} = \pi D = 2\pi r$$

(Diameter is $2 \times$ length of radius)

Examples 6.2

1. Determine the perimeter of the following figures:
   (a) A circle of 3 m radius
   (b) A rectangle 6 cm long by 5.25 cm wide
   (c) The shape in Figure 6.7.

Fig. 6.7

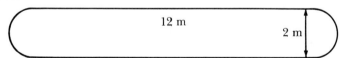

2. A perimeter fence has to be erected round a small factory whose shape is shown in Figure 6.8. What is the length of fence required?

Fig. 6.8

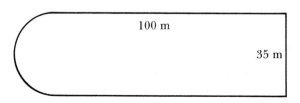

6.3 Measurement of Volumes

In the same way that squares are the rectangles with the smallest perimeters, cubes have the smallest surface area of any cuboid (rectangular block).

Rectangular blocks

$$\text{Volume} = \text{length} \times \text{breadth} \times \text{height}$$

Spheres

$$\text{Volume} = \tfrac{4}{3}\pi \times (\text{radius})^3$$

Cylinders

$$\text{Volume} = \pi \times (\text{radius of end})^2 \times \text{height} \quad (\text{see Figure 6.9})$$

Fig. 6.9

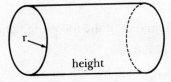

Worked example

In a small chemical factory the owner wishes to double the volume of product formed in a year. This is most easily done by doubling the volume of the cylindrical plant. Land is very expensive. It would be too expensive to make the ground cover bigger. What is the best way of doing this?

Answer

Volume of the plant $= \pi \times (\text{radius of end})^2 \times \text{height}$.

The only thing that can change is the height.

Doubling the height doubles the volume.

Worked example

What is the volume of a tower 50 metres high and 2 metres radius?

Answer

$$\begin{aligned}\text{Volume} &= \pi \times (2)^2 \times 50 \\ &= \pi \times 4 \times 50 \\ &= 200\ \pi\ \text{cubic metres} \\ &= 200 \times 3.142\ \text{m}^3 \\ &= 628.4\ \text{m}^3\end{aligned}$$

Examples 6.3

1. Determine the volumes of the following figures:
   (a) A sphere of radius 7 cm
   (b) A cube of length 7 m
   (c) A cylinder of height 10 m and radius 2 m.

2. Determine the cubic capacity of a cylindrical drying tower 12 m high and 3 m in radius.

Exercise 6

1. A building is designed to have a floor space of 240 m². Which dimensions will use least bricks in its construction?

2. A shelf was designed to be 30 cm deep by 150 cm long. Determine the area of the shelf.

3. In a supermarket the shelves are 50 cm apart, 30 cm deep and 150 cm long. 50 centilitre cans are supplied with a 5 cm base diameter or a 7.5 cm base diameter. Which gives the most efficient method of packing?

4. The following rooms have the same floor area:

   $$3\ \text{m} \times 8\ \text{m},\ 4\ \text{m} \times 6\ \text{m},\ 12\ \text{m} \times 2\ \text{m}$$

   Which will be the most expensive room to decorate, assuming the cost per m² of painting the walls and ceiling are the same?

Chapter 7

Visual Presentation of Information

In business, information often needs to be available in an immediate way. This is often done using pictures in some way. The reasons for this vary tremendously. You may want to make a sales presentation to a client, or you may wish to put over an advertisement in the media. You may simply wish to present an invoice in an easily understood form. In this chapter we are going to look at various ways in which information can be presented to give a sense of clarity and immediacy. We will look at statistical information in Chapter 9. In this chapter we will look at other information.

7.1 Tables

Information very often has to be put in the form of tables. These include accounts and invoices. They also include tables used for easy reckoning, such as Figure 5.3. Statistical tables are dealt with later (Chapter 9). Very often tables are used to put information in a sensible form. These can then be used for different purposes e.g. in drawing graphs and charts.

The most common forms of table we all know are the invoices (bills) we receive regularly, e.g. gas, electricity and rates bills. Other invoices are sent to and by firms as a method of conducting business. Let us look at some of these now.

Gas and electricity bills

These are paid quarterly and consist of two parts:
> a fixed charge for rental of the meter;
> a variable charge for the gas and electricity we use.

There are various tariffs (prices) depending on how they are used e.g. for domestic use or for customers using more than a fixed amount.

(i) Gas bills

The quarterly bills are presented in tabular form (i.e. in the form of a table).

Fig. 7.1

A typical gas bill

38.5 MJ/m³
1032 Btu's
(per cubic foot)

DATE OF READING	Meter Reading				Gas Supplied		VAT %	CHARGES £
	Present	*	Previous	*	Cubic Feet (Hundreds)	Therms		
15 May	425		407	E	18	18.576		7.06
STANDING CHARGE								8.90
VAT ON £15.96							0.00	0.00
TARIFF DOMESTIC					38.000	PENCE PER THERM		
							TOTAL AMOUNT DUE	15.96

The unit of gas is the therm where:

number of therms =

$$\frac{\text{volume of gas used (in 100s of cubic feet)} \times \text{calorific value}}{1000}$$

For example:

$$\text{number of therms used} = \frac{18 \times 1032}{1000} = 18.576$$

The number of cubic feet, the number of therms and the calorific value are all given on the bill.

114 *Visual Presentation of Information*

The number of cubic feet on the meter minus the previous reading gives the amount used in the quarter, e.g. 425 − 407E = 18.

The E means that the previous reading was estimated. This might be important on a large bill as the cost has gone up from 37.000 pence per therm last quarter to 38.000 pence this quarter. The rate is always given to 3 decimal places as it can go up by a small fraction of a penny.

The table is clear and easy to follow.

(ii) Electricity bills

The format of the electricity bill is slightly different.

Fig. 7.2   *A typical electricity bill*

Meter Readings		Units Consumed	Unit charges		VAT %	£
Previous	Present					
12 329	13 705	1376	@ 5.111 1376		0	70.33
QUARTERLY STANDING CHARGE					0	8.20
				Value excl. VAT	VAT %	
				£78.53	0	
				Amount Due		£78.53

The unit of electricity is the kilowatt hour. Otherwise the calculations are the same.

Telephone bills

Again there is a fixed rental charge. There is also a difference in meter readings to consider. However there are two extras on the telephone bill.

Fig. 7.3 *A typical telephone bill*

Rental/other standing charges		£ quarterly rate		£
From	1 Aug	System	13.45	13.45
To	31 Oct	Total	13.45	13.45
Metered units	Date	Meter reading	Units used	
	12 MAY	003571		
	11 AUG	006871		
	UNITS AT 5.00p		3300	165.00
	TOTAL (EXCLUSIVE OF VAT)			178.45
	VALUE ADDED TAX AT 15%			26.76
	TOTAL PAYABLE			205.21

There are two types of call:

>direct dialling

>operator controlled calls.

These are charged at different rates.

>VAT at 15% is charged on rental and telephone calls.

>In the above bill the total charge is £178.45.

>The VAT on £178.45 $= \dfrac{178.45 \times 15}{100} =$ £26.76

Rates bills

All properties are rated whether they are houses, shops or industrial properties. The district valuer and valuation officer of the Inland Revenue works out what you might earn by renting your property. This is based on an analysis of the size and position of your property.

It is called the gross value. He or she then assesses the possible cost of repair and maintenance. This is subtracted from the gross value to give the rateable value. This may be altered from time to time.

e.g.   Gross value   = £510
                            £113 (assessment of repairs etc)
          Rateable value   £397

Rates are then levied at so many pence in the £.

Let us assume that it is 179.75p in the £ on the above property. The rate which is levied is:

$$397 \times 179.75p = £713.61$$

The money raised by the rates is used to provide local and county services with some assistance from Government.

The 179.75p was divided up as shown:

	County council	= 175.75p
	District council	= 20.90p
	Village council	= 1.60p
Less	Domestic rate relief	= 18.50p
		179.75p

The above is simply explained in tabular form.

Other bills follow a similar pattern. Most are put on computer and this can cause problems. Look at the bill in Figure 7.4 and say what is wrong with it.

Fig. 7.4

SHEET 1   ACCOUNT No. 1038 2334 1245   11 APR 86

Department	Purchases	Cash/Credit	Date	Ref/Bill
CASH – THANK YOU		77.14	15 MAR	266–1479
FURNISHING FABRICS	70.26		24 MAR	173–6861
SOFT FURNISHING W/R	231.70		24 MAR	173–6861
FURNISHING FABRICS	328.70		24 MAR	173–6861
FURNISHING FABRICS	20.40		25 MAR	210–6884
FURNISHING FABRICS	27.00		25 MAR	211–6889
FURNISHING FABRICS	36.99		25 MAR	211–6884
SOFT FURNISHINGS W/R	57.00	25 MAR	211–6884	

Visual Presentation of Information 117

Fig. 7.5 *A simple ledger page*

1987				£	p	1987			£	p
MAY	4	Opening balance	B/d.	1264	95	MAY	7	Helliwell B/Soc	318	64
	12	Lengthy Group plc		1204	68		8	300026	162	54
	12	Inst. of Pollut. Control		684	56		14	Torquil Life Ltd.	354	65
	22	Sundries		842	61		15	Diamond Life Ltd.	164	86
				3996	80		18	300027	60	25
							18	300028	73	10
							20	300029	296	72
							21	300030	344	56
							22	Balance C/d	2221	48
									3996	80
	26	Balance	B/d.	2221	48					

On the bottom line the last two entries are in the wrong columns. Always check the accuracy of all accounts.

There are many other types of tabular statements. Two, bank statements and wage slips, will be considered here.

Accounts

A simple example of an accounts system is a bank account. This is drawn up in a *ledger*. A page of a ledger is shown in Figure 7.5.

There are two equal columns in each page, a debit (Dr) column and a credit (Cr) column. This is called Double Entry Book-keeping. Money paid to the firm is put in the debit column and money paid by the firm in the credit column. What does this mean? A cheque from a customer paid into the bank by the firm is considered to be debited from the firm to the bank. A payment by the firm to a creditor is credited to the firm by the bank. In other words the bank considers it has paid it to the firm out of its account. A bank statement is the opposite of this. This measures the money held by the bank in an account. Therefore a debit on an account is a credit on the statement. A credit on an account is a debit on the statement. This is only representing the bank situation in two different ways. All accounts are presented in the same way (see Figures 7.5 and 7.6).

Fig. 7.6

A typical bank statement
Statement of Account

1987	Sheet 48 Account No. 00	Debit	Credit	Balance Credit C Debit D	
MAY 4	Balance brought forward			1264.95	C
MAY 7	Helliwell B/Soc.	318.64		946.31	C
MAY 8	300026	162.54		783.77	C
MAY 12	Lengthy Group plc		1204.68	1988.45	C
MAY 12	Inst. of Pollut. Cont.		684.56	2673.01	C
MAY 14	Torquil Life Ltd.	354.65		2318.36	C
MAY 15	Diamond Life Ltd.	164.86		2153.50	C
MAY 18	300027	60.25		2093.25	C
MAY 18	300028	73.10		2020.15	C
MAY 20	300029	296.72		1723.43	C
MAY 21	300030	344.56		1378.87	C
MAY 22	Sundries		842.61	2221.48	C
MAY 26	Balance carried forward			2221.48	C

To show the principle clearly the ledger entry in Figure 7.5 is related to the bank statement in Figure 7.6.

Note the terms *B/d, carried forward* and *brought forward. B/d* means brought down. Sometimes *brought forward* is used. This shows the balance brought down, say, from the previous week or from the previous sheet in the ledger. In the bank statement the balance *brought forward* from the previous sheet is given at the top of the sheet and the balance to be *carried forward* to the next sheet is at the bottom.

Banks no longer keep ledgers at their offices. Instead all transactions are entered on a computer. This keeps a continuous record of money going into and out of a branch. This computer ledger is sometimes called an *audit roll*. However, the system is the same.

At the end of the day the computer ensures the 'books' are balanced. This is known as *close of business* auditing. However, banking now goes on outside normal hours (e.g. through cash dispensers). Because of this some banks maintain a continuous check on accounts. This is known as *real-time* banking.

Wages

The wages of an employee in a firm are normally based on the number of hours worked. If an employee receives a fixed amount not based on an hourly rate this is called a salary. A typical salary slip is shown in Figure 7.7.

Fig. 7.7 *Salary slip*

Item description	Rate	Amount				
Basic Pay		1541.67	Taxable Pay This Emp			13181.28
Travel/Subsistence		37.56	Tax This Emp			3272.29
Vehicle Allowance		67.34	Superannuation			693.75
	Total Pay	1646.57	National Insurance			1000.35
PAYE Income Tax		477.08				
Superannuation		77.08				
National Insurance		111.15				
Total	Deductions	665.31				
Month Ending 31 Jan 1987	Net Pay	981.26	Supercode	NI Code	NI Number	Tax Code
			Name P T Robells			

For personnel, the numbers of hours worked per day are recorded. This can be done using a clock card (Figure 7.8). From this the weekly wage is determined.

Fig. 7.8

Clock card

Clock number:		Name:					
Week ending							
	Sun	Mon	Tues	Wed	Thurs	Fri	Sat
Start							
Finish							
Start							
Finish							
Start							
Finish							
Total hours worked							

Worked example

The clock card for an employee shows that she worked the following hours:

Clock number: 35		Name: J Cadogan					
Week ending: 25 October							
Day	Sun	Mon	Tues	Wed	Thurs	Fri	Sat
Start	9.00	9.00	9.00	9.00	9.00	9.00	9.00
Finish	12.00	12.00	12.00	12.00	12.00	12.00	1.00
Start		1.00	1.00	1.00	1.00	1.00	
Finish		5.00	6.30	6.30	6.30	5.00	
Total hours worked	3	7	8½	8½	8½	7	4

She is paid at £2.80 per hour for a 35 hour week. Overtime rates are time and a quarter on weekdays, time and a half on Saturdays and double time on Sundays. How much did she earn in the week ending the 25 October?

Clock cards are often computerised and wages are worked out by the computer. Bar codes (which are also used in retail stores and some libraries) are employed to identify workers. A typical bar code is given on the back of this book. It identifies the International Standard Book Number (ISBN) and the price of the book.

Answer

Weekdays	35 hours at £2.80 per hour =	£98.00
	4½ hours at £3.50 per hour (1¼) =	£15.75
Saturday	4 hours at £4.20 per hour (1½) =	£16.80
Sunday	3 hours at £5.60 per hour (2) =	£16.80
	Total wage =	£147.35

Examples 7.1

1. Complete the following tables:

   (a) Rates bill

AUTHORITY	RATE POUNDAGE	×	RATEABLE VALUE	=	AMOUNT
Axton County Council	175.75		397		697.73
Axtonbrent District Council	20.90		397		
Asherlyn			397		6.35
Less Rate Relief	18.50		397		
			397		713.61
	TOTAL DUE		£		713.61

   (b) Car insurance

Registration detail	Premium	No Clai Discount: Amount	Amount Payable
WJS 10X	£362.56	£236.06	

122 Visual Presentation of Information

(c) Gas bill

$$38.5 \text{ MJ/m}^3$$
1032 Btu's
(per cubic foot)

DATE OF READING	Meter Reading		Gas Supplied		VAT %	CHARGES £
	Present	Previous	Cubic Feet (Hundreds)	Therms		
15 Nov	3284	2865				
STANDING CHARGE						8.90
VAT ON £					0.00	0.00
TARIFF DOMESTIC			38.000	PENCE PER THERM		
					TOTAL AMOUNT DUE	

(d) Electricity bill

Meter Readings		Units Consumed	Unit charges		VAT %	£
Previous	Present					
12 243	13 896		@ 5.293		0	
QUARTERLY STANDING CHARGE					0	
				Value excl. VAT	VAT %	
					0	
					Amount Due	£96.53

(e) *Telephone bill*

Rental/other standing charges		£ quarterly rate		£
From	1 Feb	System	13.45	13.45
To	30 April	Total	13.45	13.45
Metered units	Date	Meter reading	Units used	
	10 NOV	004862		
	12 FEB	006840		
	UNITS AT 4.4p		1978	
	OPERATOR CONTROLLED CALLS			
	12 Dec	1.84		
	Lanchestor 673214			
	TOTAL OPERATOR CALLS			1.84
	TOTAL (EXCLUSIVE OF VAT)			
	VALUE ADDED TAX AT 15%			
	TOTAL PAYABLE			

2. The meter reading on a gas meter is 2804. The previous reading was 1937. Draw up a gas bill for the premises concerned which is a small clothes store.

3. John Thomas is paid £2.40 an hour for a 38 hour week. Overtime is paid at one and a third, with time and a half on Saturday and double time on Sunday.
   His working hours for the week ending 22 November are:

Sunday	9.00-1.00
Monday to Friday	9.00-1.00 and 2.00-7.00
Saturday	9.00-1.00

   Devise a clock card for him and work out his wage for the week.

7.2 Graphs and Charts

Graphs are a visual way of presenting information. They are drawn to show how one set of values varies with another.

Charts, in the business sense, are a form of graph and will be developed alongside graphs.

Sales charts

A typical sales graph or sales chart is shown in Figure 7.9.

In this chart we can see how sales of cars from a national distribution network varied throughout the year. Sales rose steadily throughout the year except during May when there was a strike and during the firm's annual holiday period in August.

These charts are valuable in that you can see at a glance what is happening to sales. Charts or graphs give us the same information as tables but do this more immediately. Indeed tables are often used to organise the information in the first place. Therefore the graph is a second stage of visual presentation.

Let us look at the sales graph again. We can see that two sets of values are used. These are given below in table form.

Month	Jan	Feb	Mar	Apr	May	June	July	Aug	Sept	Oct	Nov	Dec
Number of sales	2800	3200	3350	5000	3000	3640	3750	1860	4130	4220	4310	4560

The two sets of values, called variables, are related. One set, the months, will always vary in the same way. They are independent of other changes. This is called an *independent variable*. The sales set is related to the months, and is called the *dependent variable*.

The independent variable (the months) is marked off evenly along a horizontal straight line on squared paper called *graph paper*. In a similar way the sales values are marked off at even distances up a vertical straight line. The two lines are at right angles to each other and are called the *horizontal axis* and the *vertical axis*.

The independent variable is marked off along the horizontal axis.

The dependent variable is marked off along the vertical axis.

When marking off your scales on the axes make sure they use most of the squared paper in both directions. Also make sure that the graph is easy to read. Later you will find that you need to read graphs accurately and easily.

There are 18 large squares along the horizontal axis. Therefore you can use $1\frac{1}{2}$ for each of the twelve months.

Fig. 7.9

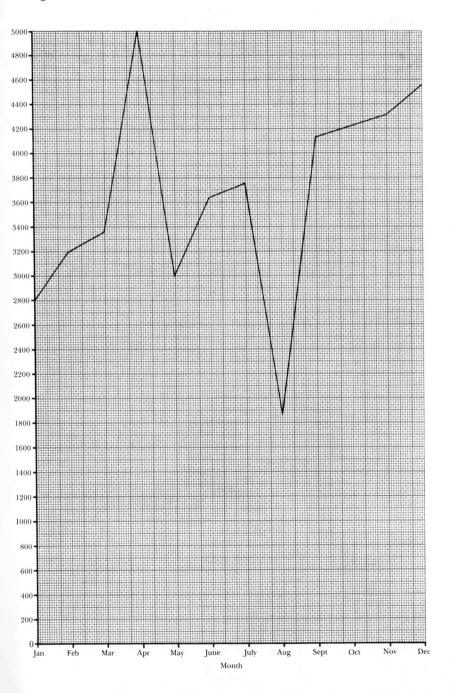

Visual Presentation of Information

There are 25 large squares along the vertical axis. The easiest scale to read, taking up most of the length is 1000 cars to 5 big squares or 100 cars to half a big square, i.e. 5 small squares.

If 5 small squares = 100 cars
1 small square = 20 cars
250 small squares = 250 × 20 cars
= 5000 cars

You must now write down what each axis represents as shown.

The vertical axis represents the *Number of cars sold*.

The horizontal axis represents the *Month*.

This is called labelling the graph. Without the labels graphs have no meaning.

Now look at Jan on the horizontal axis and go up to 2800 on the vertical axis. Mark the spot clearly with a point. Move along to Feb and go up to 3200. Mark the spot again with a point. Join the two points and continue the graph until you get Figure 7.9.

Let us try another sales chart.

Worked example

Draw a graph of the sales of Brand X soap powder from a supermarket chain during a given twelve month period from the table below. Can you think of an explanation for the unusual results of August and September?

Month	Jan	Feb	Mar	Apr	May	June	July	Aug	Sept	Oct	Nov	Dec
Number of packets sold	9265	8840	9010	9325	9300	9385	10 465	17 300	7800	8325	8740	9000

Answer

The horizontal axis is the same as for the previous example.
   For the vertical axis

the highest value = 17 255
the lowest value = 7 895
difference = 9 360

A convenient scale which takes up all the figures would be 25 squares for 10 000 packets. This gives one small square for every 40 packets sold. The vertical axis can start at 7500 and finish at 17 500 as in Figure 7.10.

Fig. 7.10

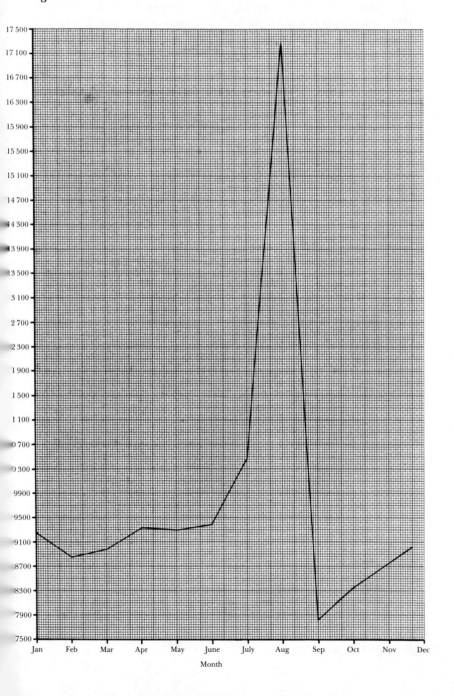

One possible reason for the large sale in August is that there was a 'give away' promotion campaign. This resulted in very large sales in August so fewer people were buying in September. Still, the overall improvement showed the campaign was a success. You may think of other reasons.

Z charts

An important type of sales chart is called the Z chart. These are used to study sales from month to month in a given year and to compare them with the previous year's sales. The best way of showing this type of chart is by a worked example.

Worked example

The sales figures for a company in 1985 and 1986 were as follows (these are given in thousands of £s to the nearest £1000):

Month	1985	1986
Jan	82	94
Feb	79	87
Mar	97	112
Apr	102	118
May	92	90
June	114	106
July	106	100
Aug	84	98
Sept	93	99
Oct	81	92
Nov	76	84
Dec	102	120

Draw a Z chart of these figures.

Answer

Draw up the table given below.

The running 12 month total is defined at the end of each month as *last* year's total sales + those for the given month this year − those for the given month last year.

Let us look at the running 12 month totals for January and February:

1985 total is 1108   Jan = 1108 + 94 − 82 = 1120
                      Feb = 1120 + 87 − 79 = 1128

The cumulative total is found for 1986 as follows

Jan = Jan figure = 94
Feb = Jan + Feb figures = 94 + 87 = 181
Mar = Jan + Feb + Mar figures = 94 + 87 + 112 = 293 and so on.

You can check if you have done this correctly. The sales figures for 1986 should equal the Dec cumulative total.

Sales for 1985–1986/(thousands of £)

Month	1986	Cumulative total	1985	Running 12 month total
Jan	94	94	82	1120
Feb	87	181	79	1128
Mar	112	293	97	1143
Apr	118	411	102	1159
May	90	501	92	1157
June	106	607	114	1149
July	100	707	106	1143
Aug	98	805	84	1157
Sept	99	904	93	1163
Oct	92	996	81	1174
Nov	84	1080	76	1182
Dec	120	1200	102	1200
	1200		1108	

Draw curves of:

(a) Running 12 month total (this shows the sales *trend*)

(b) Cumulative totals (this shows the sales *position* to date)

(c) Monthly sales for 1986 (this shows sales *fluctuations*).

Fig. 7.11

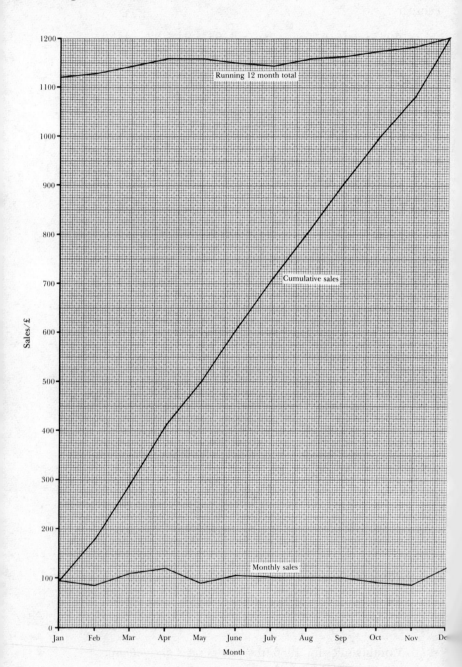

Visual Presentation of Information 131

You can easily see why this is called a Z chart.

The horizontal scale is the same as in Fig. 7.10.

The vertical scale covers 1200 units in 250 small squares or less.

A convenient scale is 2 small squares for 10 units.

Break-even charts

Another type of chart is the break-even chart.

It is extremely important to know when your sales volume will be big enough for you to break even, i.e. to begin to make pre-tax profits. Break-even charts are useful in assessing this.

However, before considering the charts, two terms need to be defined; fixed costs and variable costs.

Fixed costs are costs which are constant during the time period being considered.

Typical fixed costs are those associated with:

      depreciation (which is related to time and not to use)

      administration costs

      management costs

      rent and the energy costs of buildings.

These costs have to be met independently of the volume of sales.

Variable costs are related to sales and might have a direct proportionality or some other relationship.

Again the best way of learning about these is by using worked examples.

Worked example

A firm sells washing-machines at £350 per machine. It has fixed costs as follows:

building costs (rent, heating, lighting) = £2 900 per year

wages = £17 000 per year

advertising = £800 per year

The variable costs are £120 per machine.

Draw a break-even chart to show how many machines need to be sold to make a profit.

Answer

Fixed costs = £2900 + £17 000 + £800 = £20 700

Decide the axes as shown in Figure 7.12. Draw a horizontal straight line to represent £20 700. This does not change with the number of units. Devise a table for selling price.

Selling price:

No. of machines	20	40	60	80	100	120
Price/£	7000	14 000	21 000	28 000	35 000	42 000

Draw a graph of these points on the same paper as the fixed costs. It is a straight line (see straight line graphs Section 7.3).

Devise a table for total costs:

total costs = variable costs + £20 700

Total costs:

No. of machines	20	40	60	80	100	120
Fixed costs/£	20 700	20 700	20 700	20 700	20 700	20 700
Variable costs/£	2 400	4 800	7 200	9 600	12 000	14 400
Total costs/£	23 100	25 500	27 900	30 300	32 700	35 100

Draw a graph of these points on the same paper as the fixed costs. It is also a straight line. The last two graphs intersect where the cost price and selling price are equal.

It can be seen from the graphs that more than 90 machines must be sold in the year to allow a profit to be made.

Fig. 7.12

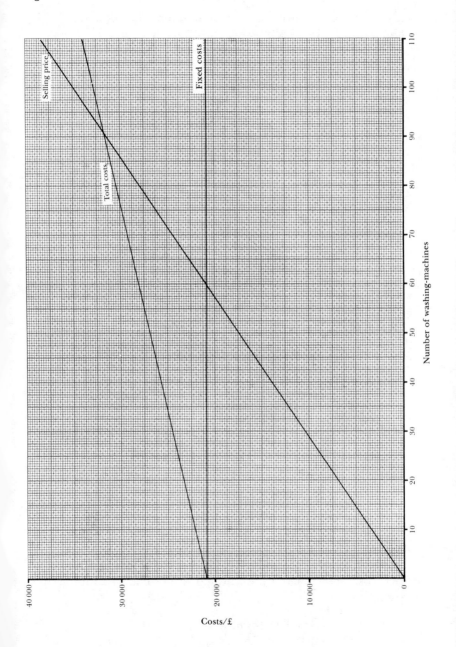

Visual Presentation of Information

Worked example

A farmer plants wheat. What yields, in tonnes per hectare, will give him a profit if he sells his wheat at £120 per tonne?

His fixed costs are £300 per hectare and his variable costs to give different yields are given below.

Yield/tonnes per hectare	1	2	3	4	5	6	7	8
Variable costs/£	15	40	80	125	185	260	390	660

At what yield will the farmer make the greatest profit?

Answer

Using the scales in Figure 7.13 draw a fixed cost horizontal line at £300.

Selling price:

Yield/tonnes per hectare	1	2	3	4	5	6	7	8	9
Selling price/£	120	240	360	480	600	720	840	960	1080

Draw the straight line (see Section 7.3).

Total costs:

Yield/tonnes per hectare	1	2	3	4	5	6	7	8
Fixed costs/£	300	300	300	300	300	300	300	300
Variable costs/£	15	40	80	125	185	260	390	660
Total costs/£	315	340	380	425	485	560	690	960

Draw a graph of total costs against yield.

This cuts the selling price graph at two points.

These are at yields of 3.25 tonnes per hectare and 8 tonnes per hectare. Between these two yields the farmer makes a profit.

The profit is given by the vertical distance between selling price and total costs. The greatest profit would seem to be at 6.5 tonnes per hectare.

$$\text{This profit} = £780 - £620$$
$$= £160$$

Fig. 7.13

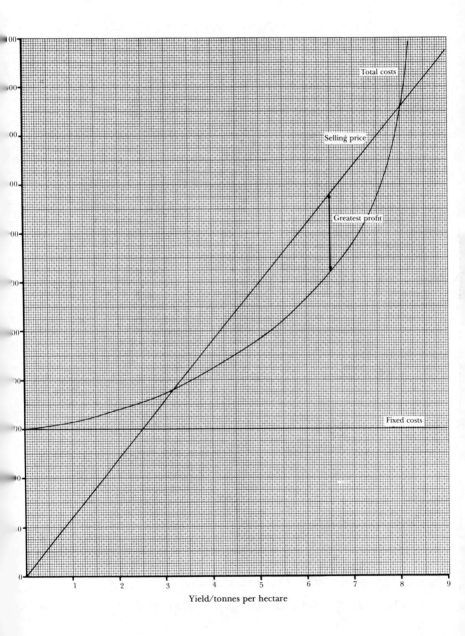

Examples 7.2

1. The monthly production of a company manufacturing motor cars was:

Month	Jan	Feb	Mar	Apr	May	June
No. of cars	83 000	85 000	86 540	87 860	89 940	83 200

Month	July	Aug	Sept	Oct	Nov	Dec
No. of cars	83 220	82 850	87 680	89 960	84 320	42 500

   Draw a production chart and suggest reasons why the numbers show sudden decreases or increases.

2. The sales figures for a pharmaceutical company in 1985 and 1986 were as follows (these are given in thousands of £s to the nearest £1000):

Month	1985	1986
Jan	56	78
Feb	82	79
Mar	91	86
Apr	83	95
May	107	114
June	91	81
July	85	99
Aug	72	59
Sept	74	81
Oct	65	80
Nov	81	75
Dec	94	102

   Draw a Z chart of the results.

3. A firm sells television sets at £304 each.
   It has fixed costs (per annum) as follows:

   Building costs (rent, heating, lighting) = £3 400

   Salaries = £34 000

   Advertising = £950

   The variable costs are £90 per machine.

   How many machines need to be sold before the firm can be assured of a profit?

4. A farmer plants barley and sells it at £155 per tonne. He has fixed costs of £310 per hectare and his variable costs to give different yields are:

Yield/tonnes per hectare	1	2	3	4	5	6	7
Variable costs/£	15	45	90	190	320	480	990

From what yields would the farmer make a profit? What would be the greatest profit made?

7.3 Straight Line Graphs

In the worked examples in Section 7.2 you saw that straight lines could be obtained. These occur when the two variables are proportional to each other. There are two kinds of proportionality – direct and indirect.

(a) When any value on the vertical axis (y-axis) goes up in a direct ratio to the related value on the horizontal axis (x-axis) the variables are directly proportional. See Figure 7.14(a).

(b) When any value on the y-axis decreases in proportion to its related value on the x-axis the variables are inversely proportional. See Figure 7.14(b).

In inverse proportionality the graph is not a straight line. However, if we plot the inverse of the horizontal *or* vertical values we do get a straight line. See Figure 7.14(c).

Fig. 7.14

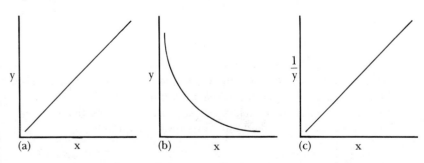

One valuable use for straight line graphs in business studies is to develop ready reckoners.

Direct proportionality

A hotel owner pays his employees £2.70 an hour but they work for periods varying from 25 to 40 hours per week. He wants a ready reckoner to work out the weekly wages of his staff. This can be done using a straight line graph. This is because the wages are proportional to the number of hours worked. In order to draw a straight line graph you need to use three points only. This cuts down the work considerably. The hotel owner will want to know the wages as accurately as possible. He will use the long axis for this if he can.

Work out the wages for the employees with the lowest and highest working hours per week. Also work out a wage near the middle.

For 25 hours, wage = 25 × £2.70 = £67.50 (lowest)

For 40 hours, wage = 40 × £2.70 = £108.00 (highest)

For 32 hours, wage = 32 × £2.70 = £86.40 (near middle)

There is roughly £40 difference between the top and bottom wage.

There is 15 hours difference between the length of the top and bottom working weeks. Plot a graph of the results (Figure 7.15).

Long axis = 25 × 1 cm units. Let 1 cm = £2. Go from £65 to £115.

Short axis = 18 × 1 cm units. Let 1 cm = 1 hr. Go from 25 to 43 hrs.

Inverse proportionality

You saw earlier that plotting graphs does not give a straight line when the proportion is inverse. See Figure 7.14(b). However if we plot the *inverse value* on the x-axis or y-axis we will get a straight line. See Figure 7.14(c).

Worked example

An asphalter employs 15 men on contract. He uses only the number he needs to do the job in the specified time. He wants to be able to rapidly assess this for each job. Devise a ready reckoner to enable him to do this. He has worked out some times given below.

No. of men	1	5	10	15
No. of hours	105	21	10.5	7

Plot a graph of the results (Figure 7.16).

Fig. 7.15

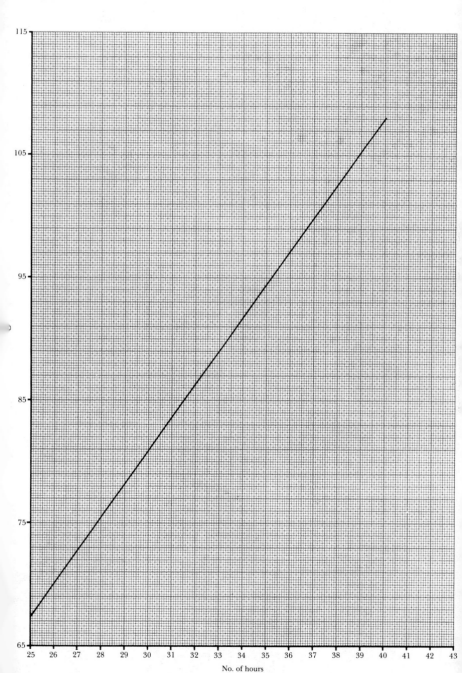

Fig. 7.16

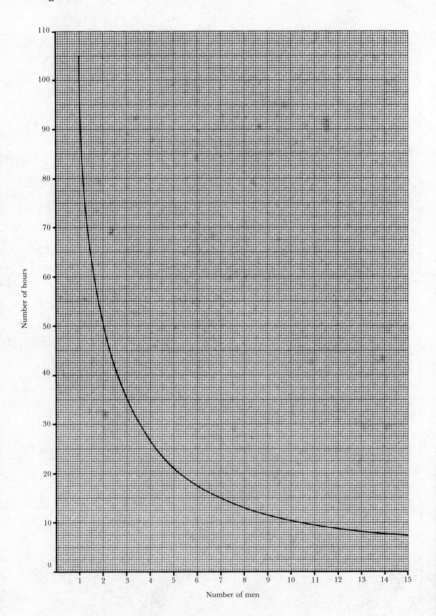

Note that the graph does not go beyond 15 men because we are not interested beyond this value. Clearly it cannot go less than 1 man. Never use values that do not make sense or that will not be useful.

Sometimes these graphs are inaccurate, or difficult to read. A straight line is frequently better. The shape in Figure 7.16 is due to inverse proportionality so it is useful to invert one set of values.

In this case the asphalter needs to know how many men to use. Therefore we invert hours. Why? If we invert men we have to invert back later because our graph gives 1 ÷ number of men.

Use $1/x$ on your calculator for inversion (Figure 7.17).

Fig. 7.17

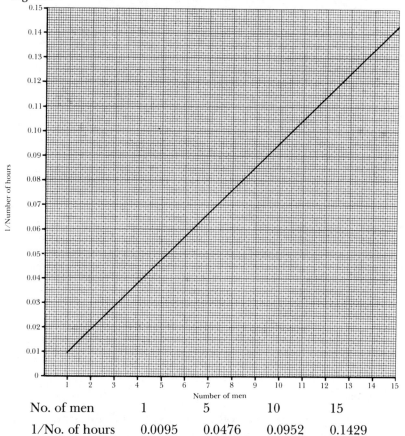

No. of men	1	5	10	15
1/No. of hours	0.0095	0.0476	0.0952	0.1429

The problem here is that you need to change hours to inverse hours which may not be convenient although using $1/x$ on your calculator is easy. A ready reckoner of this kind would be more valuable when a large workforce is involved.

142 Visual Presentation of Information

Examples 7.3

1. Devise ready reckoners for the following:

    (a) Conversion of pounds to dollars if £1 = 1.25 dollars. (This can be done rapidly for any conversion rate.)

    (b) A wholesaler sells washing-machines at £320 each and his normal order from retailers is between 1 and 10 machines. Draw graphs for a 10% and a 20% discount. What can you tell from this?

2. A builder finds that one man does a job in 90 hours, 10 men do it in 42 hours and 16 men do it in 10 hours. Draw a graph which can be used as a ready reckoner for working out any combination of men and time needed to do the job. How long will 8 men take to do the job?
Note: This is not inverse proportionality. Say why.

3. A baker charges the following varying prices for her bread:

No. of loaves	1	10	20
Price/pence	40	371	680

    Draw a graph to show what she should charge for any given number of loaves. She cannot go below 26 pence which is the cost of producing each loaf, together with minimal acceptable profit. A retailer tries to compete by buying loaves from the baker at 26 pence. How many would he have to buy at a time to get them at 26p each? What *fixed* price should he charge to compete with the baker? Assume he sells all his bread.

4. A road surface dressing contractor estimates it takes a team of 5 men 40 hours to surface 1 km^2 of road. Eight men can do it in 20 hours. 32 men can do it in 4 hours. Devise a ready reckoner for the number of men necessary to do any given job in a given time. You are limited to a minimum of 3 men and a maximum of 38 men.

Exercise 7

1. Draw up a bank *statement* for February 1987 for a publishing consultant given the following figures.

    1 February   Bank balance stood at £4856.
    3 February   Consultant was paid £1000 royalty advance by Joplins Publishing Co Ltd.

5 February	A cheque was sent to Park Print for typing expenses of £232.
7 February	Editorial fee of £65 was paid by Maisie Halts Ltd.
18 February	Annual royalty fee from Corporals and Co of £865.
20 February	A cheque of £108 was sent to Park Print for sundries.
22 February	A cheque was cashed by the consultant for £800 for living expenses for February.
24 February	A cheque was made payable to the Post Office for £30.

2. A small engineering firm carried out the following transactions during the week beginning 2 March 1987. Assuming that the cash in the bank at the end of the previous week was £84 326.18, draw up the accounts for the week. The transactions of the company are given below.

2 March	A cheque for £280 was sent to Crack Components Ltd. A cheque for £3250 was received from Finished Products Ltd.
3 March	£1865 was paid out to Carnfrith Components Ltd. £3842 was received from Rottenstick Machines plc. £2462 was paid out to Addington Adhesives plc.
4 March	£1803 was received from Metalloid Products Ltd.
5 March	£4826 was received from Cockfosters Retailers plc.
6 March	£1236 in total was paid out to 12 employees as wages. £192 was paid to the assistant manager as wages.

Assume the ledger is drawn up one page per week.

3. Your company made the following profits in 1986.

Month	Jan	Feb	Mar	Apr	May	June
Profit/£	13 840	15 860	19 220	19 000	19 310	26 840

Month	July	Aug	Sept	Oct	Nov	Dec
Profit/£	14 210	14 840	21 650	28 470	30 850	31 960

You wish to present an annual sales report. Assume you sell bedtime hot drinks. Draw up diagrams which show the figures clearly and try to explain any variations.

4. A frozen food centre made the following sales in 1985 and 1986.

Month	Sales in 1985/ thousands of £s	Sales in 1986/ thousands of £s
Jan	65	72
Feb	67	84
Mar	74	70
Apr	81	75
May	73	84
June	69	58
July	62	56
Aug	59	58
Sept	88	91
Oct	71	89
Nov	84	80
Dec	83	86

(a) Draw a Z chart of the results.

(b) Draw a sales chart of each set of results on the same piece of graph paper. Try to explain the results.

5. A farmer plants a new crop on his land and hopes to be able to charge £180 per tonne. His fixed costs are £340 per hectare. His total costs are given in the following table.

Yield/tonnes per hectare	1	2	3	4	5	6	7	8	9
Total costs/£	580	660	760	870	1030	1230	1490	1790	2100

When will the farmer make a profit? At what yield will he make the greatest profit?

6. A young man wishes to go into business on his own selling gas fires. He estimates he will sell at least 70 fires a year. His fixed costs, including his salary and that of an assistant are to be £23 750 per annum. His variable costs are £160 per gas fire. If he sells the fires at £530 is he likely to make a success of the business?

7. A ready reckoner is required to determine the exchange rate for up to a hundred pounds for values of the pound between 1 dollar and 1.5 dollars. (*Hint:* Draw a straight line for the exchange rates £1 = $1 and £1 = $1.5. Use the rate of exchange for £100 to get any other value).

Visual Presentation of Information

8. (a) Draw a conversion graph for inches to centimetres (1 inch = 2.5 cm).

   (b) Draw a conversion graph for pounds to kilograms (1 lb = 0.454 kg).

   (c) Draw a conversion graph for litres to gallons (1 litre = 0.22 gallon).

   (d) Draw a conversion graph for miles to kilometres (1 mile = 1.609 km).

9. A typical building task has been assessed by timing the workforce. It is normally done by 2 people in 48 hours, by 12 people in 8 hours and by 16 people in 6 hours. Devise a ready reckoner to enable you to assess the time taken for any number of people from 1 to 20 to do the job.

Chapter 8

Averages

We shall be looking at statistics in Chapter 9. However, we need first to look at one basic idea in statistics. The idea of averages. This is a word you use regularly, probably without thought. What does it mean?

We often want to get a rough approximation of values. At other times we need to find out the range of a set of results. For example, a manufacturer is making a certain part which should be a given size. It is impossible for it always to be exactly the same size. How much bigger or smaller can it be and still do the job? In other words how much *tolerance* can the owner get away with? The fewer parts that are scrapped the greater the profit. If too many bad parts slip through, the owner's reputation suffers. What is the best balance? Firstly, all the sizes should average out to the right size. Secondly, none should be outside accepted limits or tolerances.

To show what we mean we will look at some examples.

Worked example

A salesman does a return journey from Exeter to Bath. The distance is 132 miles. He does the outward journey at a speed of 66 m.p.h. but because of roadworks the return journey is done at 44 m.p.h. What is his average speed?

Answer

It is tempting to put:

$$\frac{66 + 44}{2} = 55 \text{ m.p.h.}$$

but this is *not* the average.

$$\text{Average speed} = \frac{\text{Total distance}}{\text{Total time}}$$

The total distance $= 132 + 132$ miles

The time taken from Exeter to Bath $= \dfrac{132}{66} = 2$ hours

The time taken from Bath to Exeter $= \dfrac{132}{44} = 3$ hours

∴ The total time $= 2 + 3$ hours

∴ Average speed $= \dfrac{132 + 132}{2 + 3}$ m.p.h.

$\qquad\qquad\quad = \dfrac{264}{5}$

$\qquad\qquad\quad = 52.8$ m.p.h.

8.1 Averages

Arithmetic mean

From the above you can see that the average can be defined as:

$$\frac{\text{The sum of a set of results}}{\text{The total number of the results}}$$

This is more correctly called the *arithmetic mean*. There are other ways of measuring averages.

In business it is frequently useful to get a rough guide for something. For example, a union wants to make sure of a fair wage for its workers. So every year it works out the average wage for workers locally or nationally. It can then try to make sure the workers do not fall behind in their position over previous years. Sometimes it is useful for employers to find the average wage of employees. It makes other financial calculations simpler.

Worked example

A printing firm employs 5 staff at the following weekly wages:

£82.50, £93, £106, £111, £116

What is the average weekly wage?

148 Averages

Answer

$$\text{Average} = \frac{\text{Total of the wages}}{\text{Number of employees}}$$

$$= \frac{£82.50 + £93 + £106 + £111 + £116}{5}$$

$$= \frac{£508.50}{5} = £101.70$$

We mentioned the average size of components earlier. Let us do an example on this.

Worked example

A food packaging company packs coffee into 200 g jars. At regular intervals batches of 30 jars are weighed to make sure the average weight falls within accepted limits.

In one batch the weights are:

198 g, 201 g, 202 g, 201 g, 203 g, 201 g, 199 g, 201 g, 200 g, 199 g, 202 g, 203 g, 201 g, 199 g, 198 g, 200 g, 201 g, 201 g, 200 g, 199 g, 200 g, 200 g, 201 g, 203 g, 198 g, 199 g, 201 g, 200 g, 201 g, 200 g,

What is the average weight? If this is a true average what does this mean for the company? (Assume the company packs 500 000 jars a day and coffee is sold at £2.76 a jar.)

Answer

We would normally use a weighted average here (see page 151). For the present we will do the long method.

Total of 30 coffee weights = 6012 g

Total number of jars = 30

$$\therefore \text{ The average weight} = \frac{6012 \text{ g}}{30} = 200.4 \text{ g}$$

This does not seem to be much out. But let us look more closely.

500 000 × 200.4 g are produced each day = 100 200 000 g

If all were 200 g exactly the weight would be = 100 000 000 g

200 000 g are 'given away' each day at £2.76 for 200 g.

$$\therefore \text{ loss} = \frac{200\,000}{200} \times £2.76 \text{ (direct proportion)}$$

$$= £2760 \text{ per day}$$

This would amount to between half and three quarters of a million pounds per year.

If this average is repeated regularly the owner is bound to try and re-align the packing machine to avoid this loss.

In Chapter 5 we looked at discounted cash flow. This was a way of helping firms to decide whether expansion was worthwhile. It was complicated. There is a fairly rough-and-ready way of doing this. But it is easy. It is called the *average rate of return*.

For example, a pharmaceutical company wants to buy a new pelleting machine. The estimated life of the machine is, say, 10 years. The manufacturer works out an estimated profit for each year less depreciation. The average is then worked out. This is used to find the average rate of return.

Worked example

The cost of the new pelleting machine is £98 500.

Year	1	2	3	4	5
Estimated profit/£	9910	10 520	11 640	13 100	14 280
Depreciation/£	9850	8865	7978	7180	6462

Year	6	7	8	9	10
Estimated profit/£	15 310	16 090	16 950	17 100	17 700
Depreciation/£	5816	5235	4711	4240	3817

The scrap value at the end of 10 years is £34 346.

Determine the average return rate for these figures.

Answer

The profits, after depreciation is taken into account, are:

Year	1	2	3	4	5
Profit/£	60	1655	3662	5920	7818

Year	6	7	8	9	10
Profit/£	9494	10 855	12 239	12 860	13 883

Averages

The average profit over 10 years =

$$\frac{£60 + £1655 + £3662 + £5920 + £7818 + £9494 + £10\,855 + £12\,239 + £12\,860 + £13\,883}{10}$$

$$= \frac{£78\,446}{10} = £7844.60$$

In using this approach we divide the capital investment by 2 to compensate for depreciation over a period.

$$= \frac{£98\,500}{2} = £49\,250$$

Average rate of return $= \dfrac{\text{average profit after depreciation}}{\text{capital cost}/2}$

$$= \frac{£7844.60}{£49\,250}$$

$$= 0.16$$

Percentage rate of return = 16%.

Suppose the depreciation value is not known for each year but only the value, if any, of the machine at the end. The following method is used.

The total profit =

$$£9910 + £10\,520 + £11\,640 + £13\,100 + 14\,280 + £15\,310 + £16\,090 + £16\,950 + £17\,100 + £17\,700$$

$$= £142\,600$$

Depreciation of capital equipment = original cost − scrap value

$$= £98\,500 - £34\,346$$

$$= £64\,154$$

Total profit	= £142 600
Depreciation	= £ 64 154
Profit after depreciation	= £ 78 446
Average profit for one year	= £7844.60
Percentage of return	$= \dfrac{£7844.60}{£49\,250} \times 100$
	= 16% (as before)

Weighted average

This is best shown by example.

Worked example

In the town of Wexhampton there are 5 engineering firms. They employ the following workforces at the wages given:

Firm 1	16 staff at £108 per week
Firm 2	12 staff at £107 per week
Firm 3	35 staff at £112 per week
Firm 4	8 staff at £101 per week
Firm 5	19 staff at £106 per week

The average of the wages in these firms acts as a yardstick for salary negotiations in the district. Find the average salary.

Answer

(i) *A common* **wrong** *way of doing this is:*

$$\text{Average} = \frac{£108 + £107 + £112 + £101 + £106}{5}$$

$$= 534/5$$

$$= £106.80$$

This ignores the different numbers of people working in the five firms.

(ii) *The* **right** *way of doing it is:*

The *total* wages are:

$$16 \times £108 + 12 \times £107 + 35 \times £112 + 8 \times £101 + 19 \times £106$$

$$= £1728 + £1284 + £3920 + £808 + £2014$$

$$= £9754$$

The *total* number of people

$$= 16 + 12 + 35 + 8 + 19$$

$$= 90$$

The average wage $= \dfrac{\text{total wages}}{\text{total people}}$

$= \dfrac{£9754}{90}$

$= £108.38$

In examples where sets of values are the same (e.g. 16 at £108) the average is called a *weighted* average.

These are not the only kind of average. Other methods of averaging have been used.

Median

When some of the numbers are very different from the rest the average can become distorted. For example, in a firm where the manager, say, earns £250 a week and the 4 people under him earn £95, £100, £103, £105,

$$\text{the average wage} = \dfrac{95 + 100 + 103 + 105 + 250}{5}$$

$= 653/5 = £131$ (to nearest £)

This has very little use even as a rough-and-ready guide. The manager's salary is well beyond the rest. In these cases we sometimes put all the values in order of size and choose the middle one. It is £103 in this case. This is much more representative of the set of values. It is called the *median*.

If there is an even number of values there is no middle one.

e.g. £90, £95, £98, £102, £103, £240

Here we take the *two* centre values £98, £102.
The midpoint of these is the median.

i.e. $(£98 + £102)/2 = £100$

Mode

There is another way of overcoming the problem of extreme values distorting results. This is shown here. Look at the worked example on page 151. The firms had:

16 staff at £108 per week
12 staff at £107 per week
35 staff at £112 per week
8 staff at £101 per week
19 staff at £106 per week.

The wage earned by most people is £112. This is considered to be the most representative value. It is the *mode*. It is the result which occurs most often.

It can be seen, therefore, that there are several ways by which we can get some kind of average value. However, this is not the whole story. Let us look at the wages in two firms. They have the same average of £107 (check) but they are very different.

Firm 1   £101 £103 £105 £107 £109 £111 £113
Firm 2   £93   £98   £102 £107 £112 £116 £121

How are they different? The spread of wages in the second firm is much greater than in the first. Therefore the average wage in the first firm is closer to all the real wages. This spread of results is important.

It is very significant in dealing with *tolerances*. If the values above are for components of mm length and the tolerance is 5 mm above or below 107 mm, then

101 and 113 are discarded by Firm 1

93, 98, 116 and 121 are discarded by Firm 2.

Twice as many components are discarded by Firm 2. Therefore the loss is twice as great. See also Appendix 1.

Exercise 8

1. Over five years a firm has given percentage annual rises of

   7.9%, 8.4%, 5.5%, 5.1%, 3.8%

   What is the percentage average rise during these years?

2. A company has 5 machines of different age and design to manufacture blocks of chocolate.

   Machine A produces £430 000 blocks per hour running for
   60 hours per week.
   Machine B produces £523 000 blocks per hour running for
   80 hours per week.

Machine C produces £486 000 blocks per hour running for 80 hours per week.
Machine D produces £500 500 blocks per hour running for 70 hours per week.
Machine E produces £528 000 blocks per hour running for 65 hours per week.

Determine the mean number of blocks per hour in the factory.

3. The wages, after tax, of 28 employees for one week with overtime were:

   £102.80, £103.40, £106.70, £110.50, £109.30, £105.50, £105.50
   £106.70, £104.80, £107.20, £106.70, £104.90, £110.60, £110.00
   £103.50, £106.70, £105.80, £106.70, £103.90, £104.50, £112.60
   £106.70, £105.90, £109.30, £110.50, £106.70, £109.20, £109.60

   Determine the mean, median and mode of the wages.

4. The cost of a new computer in an office was £8500. The estimated life of the equipment was 7 years, at the end of which it would have no financial value. The manager has assessed the profit over the 7 years to be:

Year	1	2	3	4	5	6	7
Profit	£160	£320	£560	£850	£990	£1260	£1810

   Determine the average rate of return as a percentage.

5. Determine the average wage in a group of offices employing:

Office 1	35 staff at £98 per week
Office 2	40 staff at £101.50 per week
Office 3	16 staff at £104 per week
Office 4	52 staff at £96 per week
Office 5	12 staff at £102.50 per week

   What would be the median and mode values of the wages?

6. Try, using Appendix 1, to determine the standard deviations for questions 2, 3 and 5.

Chapter 9

Statistics and the Presentation of Business Data

In earlier chapters we have been able to obtain precise answers to given problems. These answers may have been limited to some extent by the accuracy of the tables or the instruments used. However, many questions cannot have precise answers. For example, a restaurant owner cannot determine exactly how many customers he is going to have on any particular evening. Nevertheless it is important that he should be able to make a reasonable *estimate* of this number. He bases his estimate on what has happened before together with any special events which may be occurring at the time. To do this he must keep a record of the number of people visiting the restaurant on a daily basis (or even for different times in the day). He must also keep a record of the meals they eat.

The collection and logical use of this type of information is called *statistics*. As in other branches of mathematics there are sensible ways of doing this. In this chapter we are concerned with the presentation of statistical information and other data in a form which can be easily and immediately understood.

Businessmen often have to collect large amounts of figures and to relate them in a sensible way. These figures are then put in a form which can be explained easily to other businessmen or to the general public. Often the best way is a presentation in the form of tables or pictures.

9.1 Tabular Presentation

Information is very much easier to see and understand in the form of a table than when it is written down. However, to ensure that this is so, the information must be presented clearly and without any possibility of its being misunderstood. Care must be taken to make absolutely sure that the table is not so large or so complicated that it is difficult to pick out the details you want. For example, the tables drawn up below, although of very different lengths, are both easy to understand.

Fig. 9.1
Forecast demand for hotel[a] accommodation in Great Britain, 1973–85, millions of bednights

Component of demand	1973	1976			1980			1985		
		Low	Most likely	High	Low	Most likely	High	Low	Most likely	High
Domestic tourists										
Non-business	89.3	89.1	91.3	95.2	90.8	97.9	105.3	97.0	109.2	115.8
Business and conference	28.7	28.7	29.5	30.6	31.2	33.3	35.8	37.0	41.3	45.7
Total	118.0	117.8	120.8	125.8	122.0	131.2	144.1	134.0	150.5	161.5
Overseas tourists										
Non-business	47.4	49.8	53.2	57.8	55.7	60.9	69.7	65.6	76.5	90.6
Business and conference	7.6	8.2	9.3	10.9	9.2	11.2	14.8	11.6	14.1	19.7
Total	55.0	58.0	62.5	68.7	64.9	72.1	84.5	77.2	90.6	110.3
All tourists	173.0	175.8	183.3	194.5	186.9	203.3	225.6	211.2	241.1	271.8

Sources: 1973 – BHTS and HPS
1976–85 – Business and Economic Planning

[a] Licensed hotels, unlicensed hotels, boarding houses and guest houses.

Fig. 9.2
Staff and other costs for the National Sulphuric Acid Association (1985).

The National Sulphuric Acid Association Limited

Notes to the Financial Statements (continued)

Year ended 30th June 1986

7. PARTICULARS OF STAFF

	Average number of persons employed	1985
Administration	9	9

	£	£
8. PENSIONS AND WELFARE EXPENSES		
This includes a special injection to the Pension Fund for augmentation of existing pensions	£13,381	£9,138
9. TAXATION		
Tax credit relating to income from Listed Investments	(331)	—
Corporation Tax adjustment in respect of previous years	145	40
	£(186)	£40
10. STAFF COSTS		
Salaries	127,900	117,749
Social Security Costs	9,448	6,511
Other Pension Costs	21,919	30,512
	£159,267	£154,772

One employee received remuneration in the range of £35,001 to £40,000 (1985: £35,001 to £40,000).

11. LEASE COMMITMENTS
    Land and Buildings

    Annual commitment under operating leases expiring:

After five years	£47,500	£47,500

Reproduced by kind permission of the National Sulphuric Acid Association Ltd. from their 1986 Annual Report.

Sometimes the data presented are in a form which is difficult to use. For example, a restaurant served 35 customers during the course of an evening and the costs of the individual meals are given below. The table was arrived at by writing down the costs on a piece of paper as the meals were charged. The prices are in pounds.

4.00	3.50	4.70	5.80	4.50
6.05	4.20	5.00	4.15	5.25
5.10	5.50	3.75	6.00	5.10
4.80	4.50	5.30	5.50	4.60
5.00	5.05	4.55	5.15	5.00
4.60	6.40	5.75	5.90	4.40
5.60	5.40	4.60	5.15	4.55

It is difficult to make sense of these figures in this form.

To obtain a clearer picture of how the costs were spread, the table was rewritten in the following form. It can be seen that the prices range from £3.50 to £6.40, i.e. the *range* is £2.90. We now split the costs into small groups or *classes*. In this case we will use £3.50–£3.99, £4.00–£4.49, £4.50–£4.99 … £6.00–£6.49. The size of the group is chosen to make sure that it is small enough to give sensible information but not so small as to become unmanageable or meaningless. The ability to choose useful group sizes comes with experience. Group sizes, e.g. of 50p (£3.50–£3.99), are known as *class intervals*.

Having chosen the group size we can then arrange the figures (data) as shown below, putting a stroke for each value lying within any given group or class. The fifth stroke is always a crossed stroke to make the table easier to read.

Cost/£	No. of diners
3.50–3.99	1 1
4.00–4.49	1 1 1 1
4.50–4.99	⋆⋆⋆⋆ 1 1 1 1
5.00–5.49	⋆⋆⋆⋆ ⋆⋆⋆⋆ 1
5.50–5.99	⋆⋆⋆⋆ 1
6.00–6.49	1 1 1

The spread of the cost of meals is now much clearer. The above table is known as a *tally* diagram. The original values were said to be *ungrouped*. In the tally diagram they have been grouped into *definite classses.* The sizes of the groups in the above example are the same but this need not always be true (page 171). The results can now be put in another form of table. In this table the number of strokes is put in place of the actual strokes. This is even easier to understand. It is called a *frequency distribution.* The number in a given group is known as the *frequency* of the group.

Cost/£	No. of diners
3.50–3.99	2
4.00–4.49	4
4.50–4.99	9
5.00–5.49	11
5.50–5.99	6
6.00–6.49	3

As we shall see later business statistics are often expressed as percentages and the frequencies may be given as *relative frequency percentages.*

Relative frequency percentage is defined as shown.

$$\text{Relative frequency percentage} = \frac{\text{actual frequency in a class interval}}{\text{total frequency}} \times 100$$

In the above example there are two diners in the group £3.50–£3.99 and the total number of diners is 35.

$\therefore$ Relative frequency percentage $= \frac{2}{35} \times 100$

$\phantom{\therefore \text{Relative frequency percentage}} = 5.7\%$ (to the first decimal place).

If we repeat this for each set of diners we get the table below.

Cost/£	No. of diners	Relative frequency percentage
3.50–3.99	2	5.7
4.00–4.49	4	11.4
4.50–4.99	9	25.7
5.00–5.49	11	31.4
5.50–5.99	6	17.1
6.00–6.49	3	8.6
	35	99.9

(Note that the total of the relative frequencies is not exactly 100%. This is because we have calculated the percentages to one decimal place only.)

Worked example

A small town contains a factory with a total staff of 53 people between the ages of 15 and 62. Their ages are given below:

15	25	37	20	35	48	54	50	30	29	37
20	46	30	33	25	17	18	34	49	38	43
33	55	35	41	37	35	42	27	42	27	39
40	49	44	39	26	32	21	36	52	34	
31	22	32	62	28	41	36	38	58	24	

State the range of ages and express them as a tally diagram, a frequency distribution and a relative frequency percentage distribution.

Answer

The results are divided into group intervals of 15–19, 20–24, 25–29, 30–34 ... 60–64 years of age.

Since the ages range from 15 to 62 years, the range is 47 years.

Age groups	Tally diagram	Number of staff Frequency distribution	Relative frequency percentage
15–19	1 1 1	3	5.7
20–24	⊥⊥⊥⊥⊤	5	9.4
25–29	⊥⊥⊥⊥⊤ 1 1	7	13.2
30–34	⊥⊥⊥⊥⊤ 1 1 1 1	9	17.0
35–39	⊥⊥⊥⊥⊤ ⊥⊥⊥⊥⊤ 1 1	12	22.6
40–44	⊥⊥⊥⊥⊤ 1 1	7	13.2
45–49	1 1 1 1	4	7.5
50–54	1 1 1	3	5.7
55–59	1 1	2	3.8
60–64	1	1	1.9
		53	100.0

Statistics and the Presentation of Business Data 161

In the above example we have taken the ages to the nearest year. For example, a person of 35 can have any age between 34 years and 6 months and 35 years and 6 months. Variables which can have any value between given points (e.g. from 34 to 35 years) are said to be *continuous*.

However, in the previous example the costs of the meal are in a given number of pence. A meal may cost £3.99 or £4.00 but cannot have a cost lying between these values. When the variables consist of definite values they are said to be *discrete*.

Examples 9.1

1. On buying a hotel you want to estimate the cash flow for the coming year. This is produced in Figure 9.3.
   (a) In which month will you go into profit?
   (b) What is your estimated profit for the year?
   (c) Do you think Figure 9.3 as presented, is easy to read?

2. The ages of the 50 staff employed by an engineering company are as follows:

21	41	30	31	19	26	17	55	44	59
44	23	55	64	28	30	65	33	23	52
51	47	40	19	35	41	24	68	42	39
46	32	54	46	43	50	52	56	58	53
22	38	41	30	57	23	48	41	35	46

   State the range of ages and express them as a tally diagram, a frequency distribution and a relative frequency distribution.

3. The number of items sold daily by an electrical shop during the month of November is as follows:

8	16	9	3	8	9
4	7	15	5	13	7
10	4	20	11	10	11
17	22	6	7	8	19
6	14	8	10	5	8

   Separate the results into suitable class intervals and by using a tally diagram draw up a frequency distribution and a relative frequency distribution of the sales.

Fig. 9.3

Estimate of cash flow

	Jan. £	Feb. £	Mar. £	Apr. £	May £	June £	July £	Aug. £	Sept. £	Oct. £	Nov. £	Dec. £	Total £
Cash Receipts													
Sales-rooms			492	1 582	2 651	3 950	5 672	6 238	4 218	2 033			26 836
food			510	1 535	2 857	3 547	4 304	5 151	3 849	2 543			24 296
liquor			223	671	1 177	1 571	2 094	2 378	1 685	965			10 764
			1 225	3 788	6 685	9 068	12 070	13 767	9 752	5 541			61 896
Cash Payments													
Cost of sales-food			204	614	1 143	1 419	1 722	2 060	1 540	1 017			9 719
liquor			67	201	353	471	628	713	506	290			3 229
Wages and salaries	500	500	1 385	1 385	1 385	3 025	3 025	3 025	1 385	1 385	500	500	18 000
Rates					750					750			1 500
Insurance	250						250						500
Advertising	400	250	250	100									1 000
Heating and lighting		750			750			500			500		2 500
Cleaning materials		100		25	50	75	75	75	50	25			500
Repairs and replacements			1 000			100	100	100			200		1 500
Laundry		50	100	200	200	350	350	450	200	100			2 000
Administration	300	300	500	500	500	700	700	700	500	500	500	300	6 000
Loan-interest and capital	2 063			2 062			2 063			2 062			8 250
	3 513	1 950	3 531	5 087	5 131	6 140	8 913	7 623	4 181	6 129	1 700	800	54 698
Monthly surplus/deficit	*3 513*	*1 950*	*2 306*	*1 299*	1 554	2 928	3 157	6 144	5 571	588	*1 700*	*800*	7 198
Cumulative surplus/deficit	*3 513*	*5 463*	*7 769*	*9 068*	*7 514*	*4 586*	*1 429*	4 715	10 268	9 698	7 998	7 198	

Figures in italics denote overdrawn balances

9.2 Pictorial Presentation

Many people find it difficult to make sense of values even when given in tabular form. For this reason other ways have been devised to express figures in the form of pictures or diagrams which are immediately meaningful. Different types of pictures are used to explain different kinds of information to different kinds of audiences. Results involving percentages are usually expressed as:

(a) pie charts
(b) bar charts.

Bar charts are, however, not restricted to a consideration of percentages. This will be shown later (page 167).

Pie charts

These are circular and represent a pie in appearance. The whole pie is equal to 100% and appropriate slices of the pie correspond to percentage components.

The circle has an angle of 360° at the centre. Therefore, a value of 50% is represented as half the pie with an angle at the centre of 180°, i.e.

$$\frac{50}{100} \times 360°$$

The use of pie charts is best shown by an example.

Worked example

The tipping habits of hotels in the UK are shown for a given year (A) and for five years later (B).

	Year A %	Year B %
Fixed market rate for labour	3	1
No service charge/tips expected	38	24
Service charge/tips not expected	24	37
Service charge/tips expected	30	38
Incomplete reply	5	0

Draw pie charts to show these results.

164 Statistics and the Presentation of Business Data

Answer

Since the whole circle represents 100% the angle of 360° at the centre is equal to 100%. For Year A therefore:

$$3\% = \frac{3}{100} \times 360 = 11°$$

$$38\% = \frac{38}{100} \times 360 = 137°$$

$$24\% = \frac{24}{100} \times 360 = 86°$$

$$30\% = \frac{30}{100} \times 360 = 108°$$

$$5\% = \frac{5}{100} \times 360 = \underline{18°}$$

$$\underline{360°}$$

Fig. 9.4   **Fig. 9.5**

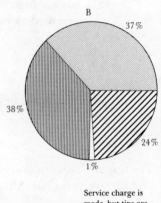

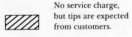

Fixed market rate for labour.

No service charge, but tips are expected from customers.

Service charge is made, but tips are not expected from customers.

Service charge is made and tips may still be expected from cstomers

Incomplete reply.

Construct arcs of a circle with angles at the centre equal to those calculated above. Use a protractor to do this (see Appendix 2). The results are shown in Figure 9.4. Figure 9.5 is constructed in a similar fashion for year B, i.e.

$$
\begin{aligned}
1\% &= 4° \\
24\% &= 86° \\
37\% &= 133° \\
38\% &= 137° \\
\hline
&360°
\end{aligned}
$$

The two charts show more dramatically than figures can how tipping habits have changed in the five years from A to B.

Bar charts

The above results can be presented in the form of a series of bars running vertically, a *vertical bar chart* (Figure 9.6), or horizontally, a *horizontal bar chart* (Figure 9.7). The *length* of the bar represents the value being measured.

Fig. 9.6

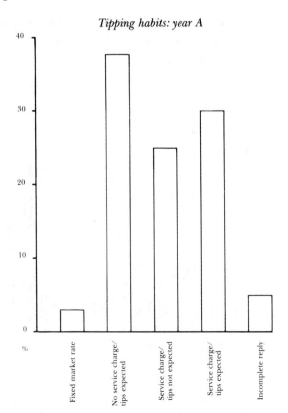

Tipping habits: year A

Fig. 9.7

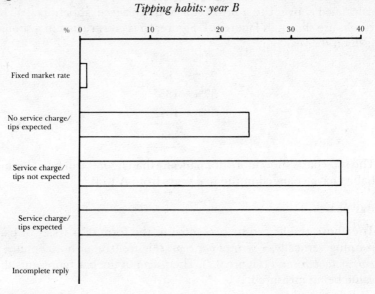

Tipping habits: year B

Bar charts are not only used to represent relative frequencies and percentages, as is shown in the following example.

Fig. 9.8

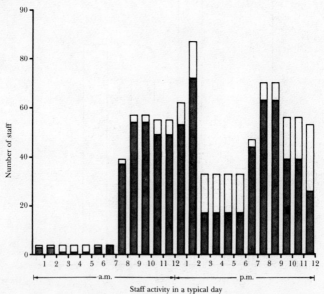

Staff activity in a typical day

Worked example

In the bar chart shown in Figure 9.8 the height of each bar represents the total number of staff on duty in a hotel during each hour of a typical day. The shaded part represents the working time and the unshaded part the idle or waiting time during each hour of the day. What fraction of the whole is the waiting time? What use may be made of these figures?

Answer

It can be seen from the diagram that the idle and working times are as follows (in terms of staff numbers).

Time (a.m.)	0-1	1-2	2-3	3-4	4-5	5-6	6-7	7-8	8-9	9-10	10-11	11-12
Idle staff	1	1	3	3	3	1	0	2	3	3	7	7
Working staff	3	3	1	1	1	3	4	38	54	54	48	48

Time (p.m.)	12-1	1-2	2-3	3-4	4-5	5-6	6-7	7-8	8-9	9-10	10-11	11-12
Idle staff		9	14	15	15	15	15	3	7	7	18	18	27
Working staff	53	74	18	18	18	18	44	63	63	39	39	27

$\therefore$ Total idle staff hours = 197

Total working staff hours = 732

$\therefore \dfrac{\text{Total idle staff hours}}{\text{Total staff hours}} = \dfrac{197}{929} = 0.21 = 21\%$

These figures may be used to cut down staff. It can be shown that with closer supervision and more efficient organisation savings of up to 12½% in staff wages can be made. Note that this is just over half the possible savings suggested by the overall value of 21%. This suggests that savings must be made on the basis of a detailed study of the chart rather than a final figure.

Bar charts can be used for comparison, by putting two series of bar charts side by side. For example, Figure 9.9 gives a comparison of the sale of houses in different price brackets in Bristol and in London.

Fig. 9.9

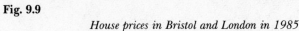

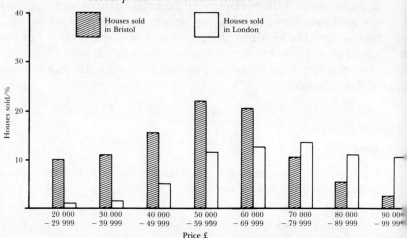

Component bar charts offer an alternative to pie charts in expressing percentages as shown in Figure 9.10.

Fig. 9.10

Part-time and full-time staff in parts of UK industry

Figure 9.10 shows how bar charts may be used to represent frequency distributions as well as percentages.

Frequency distributions are normally expressed in the form of *pictograms* or *histograms*.

Pictograms

A widely used method of presenting statistical data is to use pictures (or *pictograms*). For example, to show a number of cars, each car pictured can represent say, 10 000 cars.

Worked example

The sales of cars from Germany, Japan and the UK in a given month are:

Germany	60 000
Japan	85 000
UK	45 000

Express this information in pictogram form.

Answer

Let each picture represent 10 000 cars. The pictogram may be drawn as follows (Figure 9.11).

Fig. 9.11

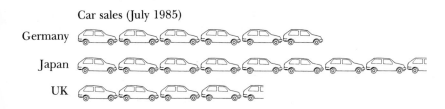

Care must be taken in drawing pictograms. The drawings of half a car must not be capable of being misunderstood.

The bar chart and the pictogram are forms of presentation with which most of us are familiar.

Histograms

The histogram is similar to the bar chart. The main difference is that the bar is replaced by a rectangle whose width is that of the class interval. The *area* of the rectangle is *proportional* to the frequency. The method of constructing histograms is shown in the following worked examples.

Worked example

The duration of conferences held in the UK in a given year were as follows:

Days	0–1	1–2	2–3	3–4	4–5	5–6
Number of conferences	77 000	11 000	7000	4000	3000	2000

Construct a histogram of these results.

Answer

Let the *horizontal* axis represent the *class interval* (days).

Let the *vertical* axis represent the number of *readings* (number of conferences).

Draw the histogram as shown in Figure 9.12.

Since the base of each rectangle is of the same length, the *height* of each rectangle is proportional to the frequency. The values 0, 1, 2, 3 days, etc. are known as the *end-points* of the class intervals and 0.5, 1.5, 2.5, etc. as the *mid-points* or *central-points* of the class intervals.

Histograms are different from bar charts. In bar charts the rectangles need not be joined (Figures 9.7–9.10). It is the heights which are important. Bar charts are used to relate discrete values. In histograms the rectangles are always joined (Figures 9.12–9.13). They are used to relate continuous values. It is the areas, not heights, which give us our measurements.

Note: Strictly speaking the values in days should read for

    0–1 days   *up to* 1 day

    1–2 days   *greater* than 1 day *but no more than* 2 days

    2–3 days   *greater* than 2 days *but no more than* 3 days.

Fig. 9.12

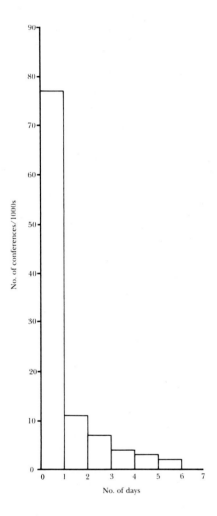

It is often useful to construct histograms with *unequal* class intervals. Since *areas* of rectangles are proportional to the frequency, changes in class intervals (length of base of rectangle) must be allowed for by changes in the vertical scale (height of rectangle). For example, if a class interval is doubled the height of the rectangle should be halved. The heights of the rectangles are no longer proportional to the class frequency.

172 Statistics and the Presentation of Business Data

Worked example

The attendance at business conferences in the UK in a given year is as follows:

Number attending	1–25	26–50	51–100	100–200
Number of conferences	58 000	30 000	9000	3000

Construct a histogram of these figures.

Fig. 9.13

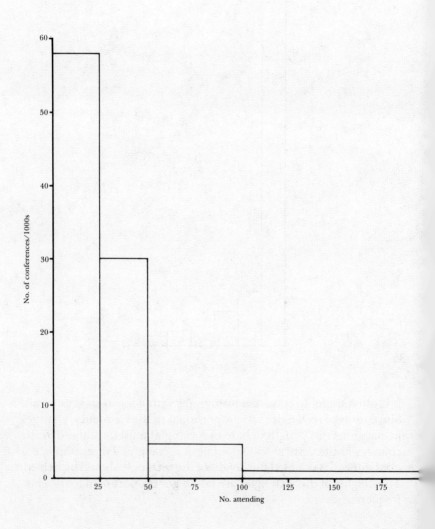

Answer

The first class interval is 25. Let height be 58 000.

The second class interval is 25. The height is 30 000.

The third class interval is 50,

i.e. 2 × first interval ∴ the height is ½ × 9000

$$= 4500$$

The fourth class interval is 100,

i.e. 4 × first interval ∴ the height is ¼ × 3000

$$= 750$$

The histogram is constructed as in Figure 9.13.

Frequently, the horizontal axis is a time axis. The histogram is then sometimes known as a *historigram*.

Examples 9.2

1. The tipping practice in the restaurants in the UK in a given year (A) and five years later (B) are given below:

	Year A %	Year B %
Fixed market rate for labour	13	14
No service charge/tips expected	80	77
Service charge/tips not expected	4	6
Service charge/tips expected	1	3
Incomplete reply	2	—

Express the above results in the form of pie diagrams.

2. The number of visitors to Britain in 1980 was as follows:

From Irish Republic	800 000
From North America	2 750 000
From Europe	5 200 000

Express these results as a pie chart.

3. The money spent by Gwylam Council in 1985 as a fraction of each £ collected was:

Service	pence
Education	66
Police	10
Social services	10
Transport	7
Fire	3
Other	4

   Express this breakdown in the form of a pie chart.

4. The number of houses sold in Cheepstow during 1986 were as given below:

Month	Jan	Feb	Mar	Apr	May	June	July	Aug	Sept	Oct	Nov	Dec
Number of houses sold	6	14	25	42	32	29	67	74	36	10	5	4

   Construct a bar chart of these data.

5. The proportion of part-time staff employed in small firms in the UK are as follows:

Size of firm	Hours worked per week		
	1–10	11–20	21–30
Less than 25 staff	26%	42%	32%
25–74 staff	15%	46%	39%
75 or more staff	17%	40%	43%

   Construct comparative bar charts of these results.

6. Draw a series of pie diagrams for the data given in Figure 9.10. Which form of chart expresses the results most suitably?

7. Express the results of problem 3 in Examples 9.1 as (a) a pictogram, (b) a histogram.

8. Express the results of problem 2 in Examples 9.1 as a histogram.

9. The total number of staff working in a large supermarket is 48. The periods of service of the staff are:

Less than 1 year	8
Between 1 and 5 years	21

Between 5 and 10 years   13
Between 10 and 15 years   6

Express these results in a suitable diagrammatic form.

Exercise 9

1. Look at the sales charts on pages 125 and 127 and redraw them as bar charts or histograms. Give reasons for your choice.

2. A large company wants to show its customers, shareholders and staff how its sales revenue is used. Suggest a suitable way of doing this, using the following data.

Sales revenue (in £m)	£1206.5
Wages, social security, pension (in £m)	£ 193.8
Interest paid on loans (in £m)	£ 25.3
Interest received (in £m)	£ 4.2
Taxation (in £m)	£ 19.3
Dividends (in £m)	£ 21.2
Cost of sales (in £m)	£ 841.3
Operating costs (in £m)	£ 58.3
Profit after tax (in £m)	£ 43.1

3. Look at the tables on

   (a) pages 176–177

   (b) page 178

   Are they easy to read and understand?

   Can you suggest ways of improving them?

4. Look at Exercise 8, problem 2. Express the production capacity of the machines as a bar chart.

5. Look at the wages given in Exercise 8, problem 3. Divide the wages into suitable class intervals. By using a tally diagram, draw up a frequency distribution, a relative frequency distribution and a histogram.

6. The value (in £m) of exports for a major food firm are:

1980	1981	1982	1983	1984	1985	1986
63.45	52.81	61.26	64.71	70.24	73.46	74.61

   Express these data in a suitable diagrammatic form.

Consumption million tonnes — Oil

	1980	1981	1982	1983	1984	1985	Change 1985 over 1984
North America							
USA	794.1	746.0	705.5	704.9	726.8	**724.1**	− 0.4%
Canada	87.6	81.7	72.9	68.2	66.7	**67.7**	+ 1.5%
Total North America	881.7	827.7	778.4	773.1	793.5	**791.8**	− 0.2%
Western Europe							
Austria	12.2	11.0	10.5	10.1	9.8	**9.8**	− 0.2%
Belgium & Luxembourg	26.6	24.5	23.3	21.1	20.4	**20.2**	− 1.0%
Denmark	13.6	12.8	11.0	10.4	10.4	**10.8**	+ 3.8%
Finland	12.8	12.3	11.3	10.5	10.6	**10.8**	+ 1.0%
France	109.9	99.0	91.5	89.4	85.9	**83.9**	+ 2.3%
Greece	12.4	11.9	11.9	11.4	11.7	**12.0**	+ 2.5%
Iceland	0.6	0.5	0.5	0.5	0.5	**0.5**	+ 10.0%
Republic of Ireland	5.9	5.1	4.6	4.2	4.1	**3.8**	− 7.2%
Italy	97.9	95.7	90.7	89.2	84.9	**85.0**	+ 0.2%
Netherlands	38.6	35.7	31.0	29.1	28.7	**29.1**	+ 1.6%
Norway	9.3	8.7	8.3	8.3	8.6	**8.8**	+ 2.0%
Portugal	8.5	8.8	9.5	9.5	9.5	**8.2**	− 14.0%
Spain	52.2	50.4	47.8	47.8	44.8	**44.0**	− 1.8%
Sweden	25.0	21.8	21.8	18.2	16.8	**16.9**	+ 0.6%
Switzerland	12.8	11.9	11.2	12.3	11.8	**12.3**	+ 3.7%
Turkey	14.8	15.4	16.5	16.2	17.0	**18.9**	+ 11.6%
United Kingdom	80.8	74.7	75.6	72.5	89.6	**77.8**	− 13.2%
West Germany	131.1	117.6	112.2	110.8	110.6	**114.0**	+ 3.2%
Total Western Europe	665.0	617.8	589.2	571.5	575.7	**566.8**	− 1.5%
Australasia							
Australia	29.7	29.2	28.4	27.3	28.5	**27.0**	− 5.3%
New Zealand	4.2	4.0	3.9	3.8	3.9	**3.8**	− 2.9%
Total Australasia	33.9	33.2	32.3	31.1	32.4	**30.8**	− 5.0%
Japan	237.7	223.9	207.8	207.2	215.1	**201.3**	− 6.4%
Total OECD	1818.3	1702.6	1607.7	1582.9	1616.7	**1590.7**	− 1.6%
Rest of NCW							
Cyprus/Gibraltar/Malta	1.5	1.5	1.6	1.6	1.7	**1.8**	+ 8.1%
Latin America	211.8	212.6	216.5	211.0	210.0	**209.5**	− 0.2%
Middle East	82.0	89.5	94.4	96.0	99.0	**98.8**	− 0.2%
Africa	70.4	74.4	77.1	78.3	80.5	**82.8**	+ 2.8%
South Asia	40.4	43.4	45.6	46.4	50.4	**53.4**	+ 6.0%
South East Asia	116.7	118.6	114.4	114.7	114.7	**112.8**	− 1.7%
Total Rest of NCW	522.8	540.0	549.6	548.0	556.3	**559.1**	+ 0.5%
Total NCW	2341.1	2242.6	2157.3	2130.9	2173.0	**2149.8**	− 1.1%
Centrally-Planned Economies (CPEs)							
China	88.0	84.8	82.4	84.7	87.5	**87.6**	+ 0.1%
USSR	436.0	444.1	450.3	446.6	440.7	**447.7**	+ 1.6%
Others	133.5	128.0	125.4	121.5	121.8	**124.3**	+ 2.1%
Total CPEs	657.5	656.9	658.1	652.8	650.0	**659.6**	+ 1.5%
Total World	2998.6	2899.5	2815.4	2783.7	2823.0	**2809.4**	− 0.5%

Total oil consumption and primary energy consumption, major countries 1980-1985.

Statistics and the Presentation of Business Data

Consumption million tonnes oil equivalent							Primary energy	
1985 Share of total	1980	1981	1982	1983	1984	**1985**	Change 1985 over 1984	1985 Share of total
25.8%	1842.0	1801.4	1727.9	1718.9	1807.8	**1799.4**	− 0.5%	24.3%
2.4%	226.1	219.8	214.9	212.0	218.0	**226.6**	+ 3.9%	3.1%
28.2%	2068.1	2021.2	1942.8	1930.9	2025.8	**2026.0**		27.4%
0.4%	26.5	25.2	24.4	24.2	24.5	**25.5**	+ 3.5%	0.3%
0.7%	50.4	48.0	46.0	44.1	46.3	**47.0**	+ 1.5%	0.6%
0.4%	19.5	18.2	16.7	15.8	16.8	**18.5**	+ 10.3%	0.3%
0.4%	21.6	21.6	21.4	21.4	21.7	**22.6**	+ 3.7%	0.3%
3.0%	189.3	186.5	181.8	181.9	186.9	**189.3**	+ 1.4%	2.6%
0.4%	17.4	16.7	16.8	17.4	18.4	**19.0**	+ 3.1%	0.3%
	1.9	1.8	1.5	1.5	1.5	**1.6**	+ 7.0%	
0.1%	8.5	8.2	8.3	8.2	8.4	**8.4**	+ 1.1%	0.1%
3.0%	145.8	144.1	140.5	138.5	139.8	**140.7**	+ 0.8%	1.9%
1.0%	73.8	70.7	67.7	64.8	67.5	**70.0**	+ 4.0%	0.9%
0.3%	30.4	29.7	32.4	30.9	31.0	**32.2**	+ 3.7%	0.4%
0.3%	11.4	11.5	12.4	12.0	12.6	**12.0**	− 5.5%	0.2%
1.6%	76.5	76.9	75.7	79.2	78.7	**78.7**	+ 0.1%	1.1%
0.6%	41.1	40.0	39.7	38.3	39.6	**40.6**	+ 2.5%	0.5%
0.4%	26.5	26.4	25.7	26.6	25.8	**27.6**	+ 7.1%	0.4%
0.7%	24.5	25.8	27.7	28.2	30.4	**33.2**	+ 9.7%	0.5%
2.8%	203.8	196.1	193.9	194.2	192.8	**201.9**	+ 4.7%	2.7%
4.1%	270.5	258.5	249.6	251.0	260.2	**267.3**	+ 2.8%	3.6%
20.2%	1239.4	1205.9	1182.2	1178.2	1202.9	**1236.1**	+ 2.8%	16.7%
1.0%	71.8	74.2	74.9	73.5	76.6	**78.7**	+ 2.8%	1.1%
0.1%	11.4	11.5	11.5	12.6	13.1	**12.7**	− 3.3%	0.2%
1.1%	83.2	85.7	86.4	86.1	89.7	**91.4**	+ 2.0%	1.3%
7.2%	359.6	353.5	341.0	342.4	369.6	**365.3**	− 1.2%	4.9%
56.7%	**3750.3**	**3666.3**	**3552.4**	**3537.6**	**3688.0**	**3718.8**	**+ 0.8%**	**50.3%**
0.1%	1.5	1.5	1.6	1.6	1.8	**2.0**	+ 15.9%	
7.5%	339.6	346.4	358.3	361.4	371.2	**380.9**	+ 2.6%	5.1%
3.5%	117.1	126.1	130.6	133.2	144.6	**144.0**	− 0.4%	1.9%
2.9%	157.7	170.2	181.6	182.8	188.9	**195.1**	+ 3.3%	2.6%
1.9%	139.1	150.3	161.2	172.7	182.4	**196.1**	+ 7.5%	2.6%
4.0%	153.4	160.8	160.1	169.8	178.2	**186.3**	+ 4.5%	2.5%
19.9%	908.4	955.3	993.4	1021.5	1067.1	**1104.4**	+ 3.5%	14.7%
76.6%	**4658.7**	**4621.6**	**4545.8**	**4559.1**	**4755.1**	**4823.2**	**+ 1.4%**	**65.0%**
3.1%	518.0	506.3	523.1	552.9	588.5	**634.3**	+ 7.8%	8.6%
15.9%	1169.0	1200.6	1247.7	1284.8	1314.9	**1376.3**	+ 4.7%	18.6%
4.4%	560.0	545.1	553.8	556.5	568.7	**580.5**	+ 2.1%	7.8%
23.4%	2247.0	2252.0	2324.6	2394.2	2472.1	**2591.1**	+ 4.8%	35.0%
100.0%	**6905.7**	**6873.6**	**6870.4**	**6953.3**	**7227.2**	**7414.3**	**+ 2.6%**	**100.0%**

Source: Extracted from *BP Statistical Review of World Energy*, 1986, BP, London.

Energy/GDP ratios for IEA countries[1]

	1973	1978	1982	1983	1984[2]	2000	Average annual growth rates (%) 1973-83	1983-84[2]	1983-2000
Canada	1.14	1.16	1.09	1.07	1.09	1.02	−0.7	2.3	−0.3
United States	1.14	1.05	0.94	0.91	0.90	0.72	−2.2	−2.0	−1.4
North America	1.14	1.06	0.96	0.93	0.91	0.75	−2.0	−1.6	−1.2
Japan	0.70	0.61	0.50	0.49	0.50	0.38[2]	−3.4	0.8	−1.5
Australia	0.68	0.73	0.74	0.71	0.70	0.63[2]	0.4	−1.2	−0.8
New Zealand	0.69	0.80	0.79	0.82	0.85	0.81	1.6	3.8	−0.1
Pacific	0.70	0.63	0.54	0.53	0.53	0.42	−2.7	0.5	−1.4
Austria	0.66	0.62	0.57	0.57	0.57	0.51[2]	−1.5	0.9	−0.6
Belgium	0.77	0.69	0.57	0.55	0.57	0.46[2]	−3.2	3.2	−1.1
Denmark	0.51	0.48	0.40	0.37	0.37	0.27	−3.2	−0.8	−1.7
Germany	0.64	0.59	0.51	0.51	0.52	0.34	−2.2	1.2	−2.4
Greece	0.59	0.63	0.62	0.64	0.64	0.77[2]	0.8	1.3	1.1
Ireland	0.96	0.91	0.82	0.80	0.77	0.75	−1.8	−3.9	−0.4
Italy	0.69	0.64	0.57	0.58	0.57	0.53[2]	−1.8	−1.0	−0.5
Luxembourg	1.94	1.51	1.20	1.18	1.24	1.31[2]	−4.9	5.2	0.7
Netherlands	0.74	0.71	0.57	0.59	0.61	0.54[2]	−2.3	3.4	−0.6
Norway	0.76	0.71	0.66	0.69	0.70	0.62	−0.9	1.7	−0.6
Portugal	0.53	0.63	0.62	0.63	0.64	0.52	1.7	2.1	−1.1
Spain	0.57	0.64	0.61	0.60	0.61	0.53[2]	0.6	0.6	−0.8
Sweden	0.69	0.67	0.61	0.62	0.64	0.48[2]	−1.1	3.8	−1.5
Switzerland	0.41	0.43	0.41	0.43	0.42	0.39	0.6	−2.8	−0.6
Turkey	0.84	0.76	0.79	0.78	0.76	0.80	−0.8	−3.0	0.1
United Kingdom	0.93	1.85	0.76	0.74	0.73	0.66[2]	−2.3	−1.6	−0.7
Europe	0.70	0.66	0.59	0.59	0.59	0.49	−1.7	0.3	−1.0
IEA Total	0.90	0.84	0.75	0.73	0.73	0.59	−2.1	−0.5	−1.2

Notes: 1 Measured in primary energy terms (tonnes of oil equivalent per $1000 of GDP at 1975 prices and exchange rates)
2 IEA Secretariat estimates

Reproduced from International Energy Agency, *Energy policies and programmes of IEA countries: 1984 review*, 1985, p. 116, by kind permission of IEA/OECD, Paris.

Chapter 10

Assignments

In this chapter the questions are on specific business situations and will involve using several of the calculations described in this book. The aims are:

(1) to help you to use calculations in real situations;
(2) to use calculations from different parts of the book in one problem;
(3) to help you to look at problems which are reasonably complex.

Try to solve the problems without guidance first of all. Then look at the hints at the back of the book. These suggest one solution to a problem but there might well be others.

1. A company has just appointed a new salesperson who needs transport. The company has three alternatives:

   (a) to buy a car;
   (b) to allow the salesperson to use his or her own car;
   (c) to hire a car.

   Given the following data, which is the best bet for the company? Assume the salesperson will travel 30 000 miles in a year.

   (a) Cost of car　　　　　　　£6500
   　　Maintenance　　　　　　£380 per year
   　　Petrol costs　　　　　　　37.9p per litre
   　　Petrol consumption　　　46 miles per gallon
   　　Depreciation　　　　　　25% per year
   　　Tax and insurance costs　£420 per year

   (b) Mileage allowance　　　for first 1000 miles 27p per mile
   　　　　　　　　　　　　　for remainder　　　16p per mile

180 *Assignments*

    (c) Hire charge
        (including maintenance)   £100 per week
       Petrol costs   37.9p per litre
       Tax and insurance costs   £420 per year
       Garaging   £150 per year
       Petrol consumption   46 miles per gallon

2. A soap powder manufacturer decides on a coupon promotion scheme. The product sells at 86p per packet. Each packet costs 56p to produce and distribute. Two schemes are envisaged.

    Scheme A offers 20p off a packet of soap powder. At this price the company would expect to pay out on 70% of the coupons issued by them (70% redemption).

    Scheme B offers 10p off a packet and 50% redemption is expected.

Purchasers normally buy one packet every two weeks.

Of those buying for the first time 50% become regular customers. Assume 85% of redemptions are new customers. In both cases 8 million coupons are to be distributed. The printing cost is 1p per coupon. Distribution costs are 2p per coupon. Traders are paid a handling fee of 25p per hundred on the coupons redeemed.

  (a) What is the cost of scheme (A) *per coupon redeemed*?
  (b) What is the cost of scheme (B) *per coupon redeemed*?
  (c) How many regular customers can be expected from scheme (A)?
  (d) How many regular customers can be expected from scheme (B)?
  (e) How much extra sales will scheme (A) encourage in a full year?
  (f) What will be the extra profit from scheme (A) in a full year? (Hint. Do not neglect the *cost* of the promotion.)
  (g) How much extra sales will scheme (B) encourage in a full year?
  (h) What will be the extra profit from scheme (B) in a full year?
  (i) Which scheme would you go for?

3. Bronxes Coffee sells in two sizes: 100 g and 200 g.

The 100 g jar sells at £1.59.

The 200 g jar sells at £3.15.

The trade sales margin is 25%.

The jars are sent to retailers in cartons containing 24 jars. The production costs in each case are:

a carton of 100 g jars £18.60
a carton of 200 g jars £35.92.

Which carton makes more profit per jar? What further information do you need to work out actual profits?

4. You are told that:

$$\text{Sales volume} = \frac{\text{profits} + \text{overheads}}{\text{unit sales price} - \text{unit variable cost}}$$

Can you explain this formula?

In this assignment you are examining the effect of price variation on profit.

(a) Toothpaste sells at 55p a tube and the manufacturer wishes to *increase* its price to 56p. At present, sales are 850 000 tubes a year. The overheads plus profits for 850 000 tubes are £212 500. The variable cost per tube is 30p. How far can sales be allowed to drop before the profit is decreased?

(b) If the toothpaste is *reduced* to 54p what is the result?

(c) Under what conditions might you choose to decrease rather than increase prices?

5. You are introducing a new brand to your range of toiletry products and you are trying to assess its value. To do this you make a financial plan for the next ten years.

Year	Additional Sales	New Materials/ Packaging Materials	Manufacturing Expenses	Advertising/ Promotion Costs	Administration Expenses
1	£36 000	£12 000	£600	£30 000	£600
2	£42 000	£13 100	£720	£15 000	£600
3	£45 000	£15 000	£720	£16 000	£600
4	£48 000	£15 900	£720	£14 500	£600
5	£48 000	£15 950	£720	£14 500	£600
6	£47 500	£15 000	£720	£15 000	£600
7	£47 400	£14 900	£720	£14 500	£600
8	£46 600	£11 000	£700	£12 000	£600
9	£45 900	£ 9 800	£600	£ 9 000	£600
10	£45 700	£ 8 500	£600	£ 6 000	£600

New plant needs to be purchased at £20 000.

(a) The above table contains all the costs. What additional information do you think is necessary to complete the financial plan?

(b) Calculate the percentage rate of return

(c) Calculate discounted cash flow over 10 years

(d) What do these figures tell you about the value of proceeding with the new product?

6. You have decided to go into business on your own. You plan to open a greengrocers with a delivery service. In order to start you have to obtain a loan from the bank. To do this you have to cost your needs. Complete the following table. Try to be realistic. Consult a local shopkeeper if you can.

   Purchase of premises, legal fees, etc.

   Mortgage on the property

   Shopfitting (shelving, etc.)

   Equipment (till, etc.)

   Stock (1 month's supply; can be allowed one month on credit)

   Van

   Running expenses

   Living expenses

   Wages for one employee

   Assume the bank will require 15% interest.

   Try to make an assessment of profit, over the next 5 years.

   What do you expect your financial situation to be then?

7. You are taking over as owner of a 50 bedroom hotel which has been going downhill mainly because of overstaffing. Describe ways in which you might make the hotel run more efficiently and cost effectively (see Chapter 9).

8. You run a small business with 5 employees. You wish to offset their wages against the sale price of your product. The records of the employees are given below. This includes an allowance for days lost due to inefficient working.

Employee	1	2	3	4	5
Hourly rate	£2.80	£2.80	£2.80	£2.80	£4.50
Normal hours per week	38	38	38	38	40
Holidays (weeks)	3	3	3	3	4
Bank holidays (days)	10	10	10	10	10
Sickness (days)	6	9	4	1	0
Allowance for inefficiency (days)	16	16	16	16	10

(a) Determine the total number of productive hours in a year.

(b) Determine the hourly rate for productive hours.

(c) What is the average rate offset against sales price?

(d) Overheads are £14 800 per year. What would be the overheads per productive hour?

(e) A customer has a job done which takes twelve hours. Raw materials cost £78. What would be the total cost per hour for the job?

(f) What would be the *total cost* of the job?

(g) Allowing 15% profit on the cost price and 15% VAT how much would the customer be charged?

9. In assignment 8 no allowance was made for National Insurance contributions or for income tax. Statutory Sick Pay is not relevant as all sick days were single ones.

The employee's contribution to National Insurance is 9% of wages.

The employer's contribution to National Insurance is 9% of the wages of employees 1–4 and 10.45% of that of employee 5.

All employees are married men with no children. They are allowed £3455 free of tax and pay 29% on the rest.

(a) How much does each employee pay in tax and National Insurance in a week?

(b) How much does the employer pay in National Insurance contributions in a year?

(c) How much would he have to increase his costs per productive hour to allow for this?

10. You wish to assess the performance of your company. What indicators might help you to do this? Suggest ways in which you could present an accurate and understandable picture of the company's performance to your employees.

11. Apply the ideas outlined in your answers to assignment 10 to the following figures for a manufacturing company on 31 December 1986.

Assets

	Buildings and land	£185 000
	Plant and machinery	£44 000
	Fixtures and fittings	£6 000
	4 cars	£28 500
	Investments	£10 000
	Goodwill	£10 000
	Debtors	£38 000
	Stock	£31 000
	Work in progress	£6 000
Capital		£20 000
Reserves		£20 000
Profit and loss a/c		£243 000
Current liabilities		
	Bank overdraft	£26 000
	Creditors	£29 000
	Hire-purchase	£9 500
	Current taxation	£8 500
	Dividend	£2 000
Turnover		£732 000

Outgoings

Salaries/wages	£286 000
Materials	£200 000
Rates	£1 400
Electricity	£1 950
Telephone	£1 100
Insurance	£3 000
Hire-purchase	£1 900
Car expenses	£1 500
Advertising	£1 000
Bad debts	£1 100
Bank charges	£900
Depreciation	£18 000
Taxation	£44 500

12. A company has been experimenting recently with methods of wrapping chocolate bars. The original covering was 15 cm by 15 cm. This was first reduced to 14.5 cm by 14.5 cm by changing the angle of packing. It was later cut down to 15 cm by 11.7 cm. These were sizes obtained by averaging out samples of different wrappers of each kind. Assuming that the first is 100%:

    (a) What are the percentage savings using the other two sizes?
    (b) The foil is 8 micron thick and has an area of 46 m^2 per kg (1 micron = 1 millionth of a metre). What is the weight in kg of packaging for 1 000 000 bars in all three sizes?
    (c) The cost of foil is £2.50 per kg. What is the cost of packaging for 1 000 000 bars for each size of packaging?
    (d) What are the savings of the second and third sizes in £s to the nearest £?

13. A salesman has to make regular journeys between two depots in Exeter and Broxbourne. He can either go by motorway or by main road. The journey by road is 232 miles and takes on average 6¼ hours. The journey by motorway is 258 miles and takes on average 4¾ hours. He is given a mileage allowance of 28 pence per mile. He is paid a salary of £17 000 per annum. He normally works a 35 hour week and has 4 weeks holiday.

    (a) What is his hourly rate?
    (b) What is the cost of each journey by the different methods?
    (c) What is the most cost-effective method of travel?

14. A firm produces a chocolate bar to the following dimensions:

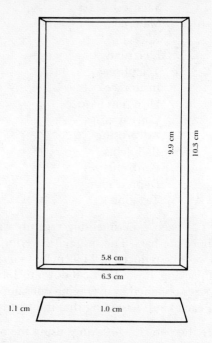

As a sales promotion for an increased size of bar (10% larger), the company sells the first 65 000 000 bars at the old price of 19p before putting them up to 21p.

(a) The increase in size can be measured in length. Why?
(b) The original bar cost 9p to make. What will the new bar cost?
(c) Estimate how much the sales promotion will cost. How many extra bars will have to be sold to compensate for the promotion?

Answers

Chapter 1

Exercise 1

1. *Assessing local company needs*
   Check how the company structures have changed in recent years. Has the number and size of companies grown or fallen? How much work has been generated? Has it grown or fallen? You then need to try to forecast the future from these results. You will see how to do this in Chapter 9.

   Staff needs
   You have decided how much work to expect. Now you need to know how much work one typist or secretary can do. How many typists are needed altogether? How many hours will be worked? Allow for holidays and sickness. You will probably be the supervisor. Chapters 2 to 5 are important here. Chapter 8 may also be useful.

   Equipment needs
   Decide what you need, e.g. typewriters, filing cabinets. How many? Allow for breakdown.

   Finance
   You need to work out possible profits as well as costs. Probably over several years. This you will present to the bank to persuade it to back you. Chapter 5 is vital here. Chapter 7 could also be important in trying to sell yourself to the bank.

2. You would need to do a similar exercise to 1. In this case you would have to work out realistic sales. The choice of boilers is important. If you have one large boiler how do you cope with maintenance? How do you allow for changes in demand and workload? On the

other hand small boilers may be originally more expensive and take up more ground space. You will have to balance advantages and costs. You will again find Chapters 2–5 and 9 important. Chapter 6 could be vital in choosing the correct size of boiler.

3. You would have to determine any deductions involved, e.g. National Insurance contributions, tax, pension. You would then decide on a form which could be easily understood by the employees. See Chapter 7 for this. The salary slip should fit in with the accounting system which may be computerised.

In all the above cases you will have to make decisions involving mathematics. Sometimes they will be simple, sometimes difficult. An understanding of business mathematics with a computer or calculator to do the drudgery is vital to success in business.

Chapter 2

Examples 2.1

1. (a) 107 (b) 145 (c) 355 (d) 575 (e) 1151
2. £3845
3. £109 494
4. 196 hours
5. £868

Examples 2.2

1. (a) 14 (b) 26 (c) 203 (d) 18 (e) 186
2. £7668
3. £6150
4. £147
5. £4907

Examples 2.3

1. (a) 1435 (b) 8022 (c) 30 912 (d) 187 124 (e) 2 244 504
2. £4424

3. 1000 miles at 28p per mile = 28 000p
   26 362 miles at 24p per mile = 632 688p
                    Total = 660 688p or £6606.88

4. 900

5. £26 100 per checkout. Total = £234 900.

Examples 2.4

1. (a) 64 (b) 185, remainder 5 (c) 364 (d) 126, remainder 10

2. £23

3. £1 45p (or 145p)

4. £3 75p (or 375p)

5. 13p

Examples 2.5

1. 33

2. 4

3. 4

4. 13

5. 73

Exercise 2

1. (a) £3514 80p (b) £2711 40p
   He should choose the first. (*Note:* as mileage increases (b) becomes more attractive.)

2. 44, £5464

3. Expenses are £160. Payment for work is £140 per day. The consultant charges £20 per hour.

4. Petrol costs = £7774
   Total costs = £16 266

5. (a) £1207 (b) £524 (c) £27 248

Chapter 3

Examples 3.1

1. (a) $3^{1}/_{24}$ (b) $^{2}/_{51}$ (c) $1^{3}/_{12}$ (d) $^{7}/_{5}$

2. A gets £25 000
   B gets £20 000
   C gets £35 000

3. £69 888 is ploughed back, £4992 for each employee, £9984 for the manager.

Examples 3.2

1. (a) 249.8 (b) 74.81 (c) 1188.06

2. (a) 14 588.034 (b) 701.23 (c) 275.23

3. £2.27

4. 290.93, £13 440.10

5. 179 400 000 grams
   180 600 000 grams

6. 0.37p (to 2 dp)

Examples 3.3

1. (a) £51.63 (b) 583.6 metres (c) 846.33 litres (d) correct (e) 64.33

2. £577.14

3. £68 200

4. 230.4p

5. £12 484

Exercise 3

1. $^{15}/_{28}$

2. $^{7}/_{18}$, £375 667 (to nearest £1)

3. £14 717.95, £14 192.31, £22 089.75

4. 0.140 (to 3 dp), £8000

5. (a) £2818.42 (b) 5.47 (to 2 dp)

Chapter 4

Examples 4.1

1. (a) 20 000 centimetres (b) 5 kilometres (c) 0.5 decametres
   (d) 380 000 millimetres (e) 98 460 centimetres

2. £148.72

3. (a) £2098 (b) £2720 (a) is best

Examples 4.2

1. (a) 0.70 litres (b) 2000 millilitres (c) 3 000 000 cubic centimetres
   (d) 8.849 cubic hectometres (e) 569 000 litres

2. A 70 cl bottle at £1.40 is equivalent to 1 litre at £2.00.
   The 1 litre bottles are the better buy.

3. £9.65, 378 miles (to nearest mile)

4. 9 miles (to nearest 0.1 mile)

Examples 4.3

1. (a) 8000 kilograms (b) 0.03 kilograms (c) 5600 grams
   (d) 7.564 tonnes (e) 8560 kilograms

2. The second wholesaler charges 50.6p per kilogram. Therefore the first wholesaler offers a slightly better buy.

3. 314 tonnes (to nearest tonne)

Exercise 4

1. 51 515, £800.87

2. Route 1 is 44.3 kilometres
   Route 2 is 53 kilometres
   Route 1 is the more profitable route

3. Route 1 £1.28
   Route 2 £1.53

4. £1.95, £557 (to nearest £1)

5. (a) £72.87 (b) £3789.24

Chapter 5

Examples 5.1

1. $^{16}/_3$, 16:3
2. $^3/_{11}$, 3:11
3. $^1/_{18}$, 1:18

Examples 5.2

1. *Actual* days worked are 228; rate to employer is £3.42 per hour.
2. The length of working day is 12.36 hours. Overtime is 4.36 hours.
3. £30
4. £168
5.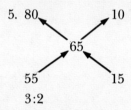
   3:2
6. £6000, £10 500, £25 500
7. Present staff   = 110 men
   On new scheme = 90 men
   20 men redundant
8. Additional cost = £98.56; cost at 13p = £1281.28; cost at 18p = £1774.08; saving = £492.80
9. Cost per day $=\dfrac{1 \times 350 \times 1 \times 1 \times 37.4}{1 \times 31 \times 22 \times 1}$

   $= £19.19$

Examples 5.3

1. Service charge = £2.25, VAT = £3.71, Total = £28.46
2. 12.7% (to one dp)
3. 78.5%
4. 71.6% (to one dp)

Answers 193

5. 10 667, £4158.62

6. $\dfrac{\text{Gross profit}}{\text{Sales}} = 46\%$

   $\dfrac{\text{Net profit}}{\text{Sales}} = 34\%$

   $\dfrac{\text{Overheads}}{\text{Sales}} = 12\%$

   Assume initial investment
   $$= (\text{sales} - \text{profit}) + \text{unsold stock} + \text{overheads}$$
   $$= £181\,600$$
   $$\text{Interest} = £19\,976$$

   The investment is much less profitable. This is supported by the very high efficiency ratios.

7. $\dfrac{\text{Current assets}}{\text{Current liabilities}} = 1.7$

   $\dfrac{\text{Current assets} - (\text{stock and work in progress})}{\text{Current liabilities}} = 0.99$

   Both values show that the liquidity of the firm is sound.

8. $\dfrac{\text{Money owed by debtors}}{\text{Annual sales}} = \dfrac{£3000}{£22\,000} \times 365 \text{ days} = 50 \text{ days} \quad (1)$

   $= \dfrac{£3800}{£29\,000} \times 365 \text{ days} = 48 \text{ days} \quad (2)$

   The trend is a healthy one as the proportion of money owed is decreasing.

9. $\dfrac{\text{Money owed to creditors}}{\text{Annual purchases}} = \dfrac{£1900}{£4500} \times 365 \text{ days} = 154 \text{ days} \quad (1)$

   $= \dfrac{£2300}{£6350} \times 365 \text{ days} = 132 \text{ days} \quad (2)$

   The trend is a healthy one for the same reason as in 8.

Examples 5.4

1. £63 654.21

2. The bakery borrows £124.50, it pays back £161.80.

3. (a) £3106.25 (b) £18 637.50

4. Profit on investment = £125 797 (to nearest £1)
∴ Business expansion appears to be the better bet.

5. Profit = £125 797 + £14 500 = £140 297
Director better off by £140 297 − £130 000 = £10 297

6. £695 024 (to nearest £1)

7. £3469 (to nearest £1)

8. £393.60

9. £5976

10. Profits (at present prices) for seven years
    = £20 694 (to nearest £1)
Money recouped in 7 but not 6 years.

11. (a) Profit = £14 768.87
    (b) Profit on shares + dividend = £6806.28 + £1282.72
        = £8089
    (c) Interest + dividend = £2178.77 + £4581.01
        = £6759.78
    (a) is the best investment

12. Mark-up   = 34% (to nearest whole number)
    Margin   = 25% (to nearest whole number)
    Mark-down   = 10% (to nearest whole number)
    New mark-up = 21% (to nearest whole number)
    New margin = 17% (to nearest whole number)

Exercise 5

1. £1276.20

2. If you increase sales by 10%, you increase the profits by 10%.
   $^{10}/_{100}$ (£83 000 − £38 800) = £4420

   If you increase prices by 3% you increase profits by 3%.

   $^{3}/_{100}$ × £83 000 = £2490

   (a) would appear to be better, but the additional sales volume may be more difficult to achieve than retaining the same volume at a higher price.

3. (a) Each worker must do 1.1 extra days per week in the 9 weeks the job takes.
   (b) Each worker must do 9 hours per week extra.
   Normal rate is £3 per hour (£120 ÷ 40).
   Overtime rate is £4.50.
   Overtime earned in 9 weeks (45 days) is $81 \times £4.50 = £364.50$.
   Total overtime earned $= 36 \times £364.50 = £13\,122$.

4. £4163.82

5. (b) £15 384.25
   Weekly interest = £295.86. She may choose (b) in the hope that with a wider range of products she may recoup the £295.86 per week and retain her independence. The reasons for choosing (a) are less responsibility and no worry over interest.

6. Mark-up = 14.3% (to one dp) (% of cost price)
   Margin = 12.5%   (% of selling price)

7. (a) £10 200

   (b) $\dfrac{\text{Gross profit}}{\text{Sales}} = 41\%$

   $\dfrac{\text{Net profit}}{\text{Sales}} = 12\%$

   $\dfrac{\text{Overheads}}{\text{Sales}} = 29\%$

   The middle value is most significant.

   (c) Stock turnover = 13%. The performance of the company is improving.

   (d) $\dfrac{\text{Current assets}}{\text{Current liabilities}} = 1.5$ (to one dp)

   $\dfrac{\text{Current assets} - (\text{stocks and work in progress})}{\text{Current liabilities}} = 1.0$

   Yes

8. (a) 93 days (b) 112 days
   The liquidity of the firm is not as good as in the previous year. Improvements in liquidity are accompanied by a decrease in the ratios.

9. Purchase price = £2945.25
   Mark-up = 34% (to nearest whole number)
   Margin = 26% (to nearest whole number)

10. (a) In the seventh year
    (b) £18 958.90

11. (a) Profit (before tax) at end of year 2 = £5452.37
       Profit (after tax) at end of year 2 = £4912.59
    (b) Dividend (year 1) = £148.44
       Dividend (year 2) = £148.44

       Profit at end of year 1 = £148.44
       Profit at end of year 2 = £5061.03

Chapter 6

Examples 6.1

1. (a) 24 m^2 (b) 52.5 cm^2 (c) 154 cm^2

2. 3 lengths 4.5 metres long, area = 13.5 cm^2

3. 962 m^2

Examples 6.2

1. (a) 18.852 m (b) 22.5 cm (c) 30.284 m

2. 290 m (to nearest whole number)

Examples 6.3

1. (a) 1437 cm^3 (to nearest whole number) (b) 343 m^3 (c) 125.68 m^3

2. 339 m^3 (to nearest whole number)

Exercise 6

1. A square building, length = breadth = 15.5 m

2. 4500 cm^2

3. Using cans with 5 cm base: height = 25.5 cm
   We can get one row of 180 cans.

Using cans with 7.5 cm base: height = 11.3 cm
We can get 4 rows each containing 80 cans = 320 cans.
The cans of 7.5 cm base give more efficient packing.

4. The room 12 m × 2 m long

Chapter 7

Examples 7.1

1. (a)

AUTHORITY	RATE POUNDAGE	√	RATEABLE VALUE	=	AMOUNT
Axton County Council	175.75		397		697.73
Axtonbrent District Council	20.90		397		82.97
Asherlyn	1.60		397		6.35
Less Rate Relief	18.50		397		73.45
	179.75		397		713.61
	TOTAL DUE		£		713.61

Note: The amount should add to 713.60. Difference due to approximations in multiplying.

(b)

Registration detail	Premium	No Claim Discount: Amount	Amount Payable
WJS 10X	£362.56	£236.06	£126.50

(c)

38.5 MJ/m³
1032 Btu's
(per cubic foot)

DATE OF READING	Meter Reading		Gas Supplied		VAT %	CHARGES £
	Present	Previous	Cubic Feet (Hundreds)	Therms		
15 Nov	3284	2865	419	432.408		164.32
STANDING CHARGE						8.90
VAT ON £					0.00	0.00
TARIFF DOMESTIC			38.000	PENCE PER THERM		
					TOTAL AMOUNT DUE	173.22

(d)

Meter Readings		Units Consumed	Unit charges			VAT %	£
Previous	Present						
12 243	13 896	1653	@ 5.293			0	87.49
QUARTERLY STANDING CHARGE						0	9.04
					Value excl. VAT	VAT %	
						0	
						Amount Due	£96.53

(e)

Rental/other standing charges		£ quarterly rate		£
From	1 Feb	System	13.45	13.45
To	30 April	Total	13.45	13.45
Metered units	Date Meter reading Units used			
	10 NOV 004862			
	12 FEB 006840			
	UNITS AT 4.4p		1978	87.03
	OPERATOR CONTROLLED CALLS			
	12 Dec 1.84			
	Lanchestor 673214			
	TOTAL OF OPERATOR CALLS			1.84
	TOTAL (EXCLUSIVE OF VAT)			102.32
	VALUE ADDED TAX AT 15%			15.35
	TOTAL PAYABLE			117.67

2. —

3. £56

Examples 7.2

1. The fall in production in June to August could have been due to summer holidays. The slide in November could be due to illness, e.g. influenza. This with holidays makes December even worse.

2. —

3. Fixed costs = £38 350; 180 sets must be sold to break even.

4. Profit lies between 2.5 and 6.8 tonnes per hectare.
   Greatest profit is £150, at 5 tonnes per hectare.

Examples 7.3

1. —

2. A straight line is obtained when 'men' is plotted against 'hours'. Remember a curve is obtained for inverse proportion (page 137). 8 men do the job in 52.5 hours.

Answers

3. Plot 'no. of loaves' against the cost of *each* loaf when bought, e.g. 20 and 680/20p.

   At 43 loaves no further savings can be made.
   A sensible charge might be halfway between 26p and 40p. You might think of a better price.

4. Plot 1/hours against the number of men from 3 to 38 men.

Exercise 7

1. —

2. —

3. There is a sudden drop in July following a large rise in June. There may have been a product promotion in June.

4. —

5. Profit lies between 3.2 and 6.6 tonnes per hectare.
   Greatest profit is £80 at 5.5 tonnes per hectare.

6. The breakeven point is 64 fires; 70 fires brings in a profit of £2150. This does not look promising. (*Note:* 100 fires gives £13 250 profit which is more acceptable.)

7. Mark off the required number of dollars for £100 on the £100 line. Join this point to the zero value. Read from the line the values of $s or £s you require at this rate of exchange.

8. —

9. Inverse proportion; plot 1/ hours against the number of men from one to 20 men.

Chapter 8

Exercise 8

1. 6.14

2. 495 400 (to nearest 100)

3. Mean   = £107.01
   Median = £106.70
   Mode   = £106.70

4. Total profit = £5950 (after depreciation)
   Average profit for one year = £850
   Average rate of return = £850/(£8500/2) = 0.2
   = 20%

5. Mean = £99.20
   Median = £98.00
   Mode = £96.00

6. Standard deviation = 33 380

7. Standard deviation = £2.50

8. Standard deviation = £2.90

Chapter 9

Examples 9.1

1. (a) August
   (b) £7198
   (c) There could be one problem. It is difficult to distinguish between italics and other numbers. It may have been easier to put bold numbering instead of italics.

2. Age range 17–68
   Suitable class intervals

10–19 years	1 1 1	3	6
20–29 years	⊬⊬⊤ 1 1 1	8	16
30–39 years	⊬⊬⊤ ⊬⊬⊤	10	20
40–49 years	⊬⊬⊤ ⊬⊬⊤ 1 1 1 1	14	28
50–59 years	⊬⊬⊤ ⊬⊬⊤ 1 1	12	24
60–69 years	1 1 1	3	6

3. | | | | |
   |---|---|---|---|
   | 1–5 | 1 1 1 1 | 5 | 16.7 |
   | 6–10 | ⊬⊬⊤ ⊬⊬⊤ ⊬⊬⊤ | 15 | 50 |
   | 11–15 | ⊬⊬⊤ | 5 | 16.7 |
   | 16–20 | 1 1 1 1 | 4 | 13.3 |
   | 21–25 | 1 | 1 | 3.3 |

Examples 9.2

1. —
2. —
3. —
4. —
5. —
6. This is a matter of preference. I prefer Figure 9.7.
7. —
8. —
9. —

Exercise 9

1. —
2. A bar chart and pie diagram are equally appropriate in presenting this information.
3. (a) and (b) are both tables on the consumption of energy.
   (a) This table is restricted to the consumption of oil compared with primary energy. It is large but very easy to understand.
   (b) This is not easy to understand unless you know what GDP is (the gross domestic product for the world nation by nation). You must also appreciate that the use of negative growth rates means less energy is being used. The table measures changes in the efficient use of energy.
4. —
5.

102–103.99	1 1 1 1	4	14.3
104–105.99	L̶H̶T̶ 1 1	7	25.0
106–107.99	L̶H̶T̶ 1 1 1	8	28.6
108–109.99	1 1 1 1	4	14.3
110–111.99	1 1 1 1	4	14.3
112–113.99	1	1	3.6

6. A bar chart is appropriate here.

Chapter 10

1. (a) You must take into account the fixed costs. Include some allowance for interest by investing £6500.
   (c) You must take into account the fixed costs.

   The running costs in (a) and (c) are the same.

   The fixed costs have been absorbed in (b) and would be paid by the salesperson.

   The cost of petrol for 30 000 miles is £1124 (to nearest £1).

   The mileage allowance for 30 000 miles is £4910 (to nearest £1).

   Compare the difference between these and fixed costs to reach your decision.

2. In this exercise you are *estimating* figures (e.g. 70% redemption). The results of calculations (a) to (h) should help you to reach the most logical solution in (i).
   (a) 22.75p
   (b) 16.25p
   (c) 2 380 000
   (d) 1 700 000
   (e) 61 880 000
   (f) £17 290 000
   (g) 44 200 000
   (h) £12 610 000
   (i) Consider factors like differences in outlay. What happens if redemption estimates are wrong? What happens if fewer customers become regular?

3. Retailers make more profit on a 200 g jar (78p) than on a 100 g jar (40p) but not weight for weight. Production costs are only part of the costs. However, on this basis the producer makes more money per 200 g jar, than on a weight for weight basis. The picture is not complete without costs like distribution costs.

4. (a) Sales may drop to 817 308 tubes per year.
   (b) If profits and overheads remain the same the volume of sales must increase to 885 417 (4% increase in sales).

5. (a) You need to include yearly profits and cash-flow, taking into account capital costs.

(b) See page 81 for method.

(c) See page 93 for method.

(d) When making your decision consider the *merits* of (b) and (c).

6. The previous question has given some hints on how to plan and prepare a forecast. Also Figure 9.3 may give you some ideas.

7. You would tabulate the staff hours throughout a working day, and possibly a working week, allowing for holidays and illness. Consideration of idle time and working time would give an impression of overall efficiency. Look at Figure 9.8 and see what savings you can make in this case.

8. (a) 6928 (men) + 1830 (manager) = 8758 hours (productive)

    (b) £3.19 (men); £5.11 (manager)

    (c) Average = £3.57

    (d) Overheads per productive hour = £1.69

    (e) £11.76

    (f) £141.12

    (g) £186.63

9. (a) 1–4 (men) = £21.17; 5 (manager) = £49.13

    (b) £2969.92

    (c) 34p per productive hour

10. For performance indicators see page 80.

    You would probably want to use a combination of visual methods for a presentation to employees by talk, exhibition and/or letters. This could include sales figures, estimates of profit and/or a simple set of accounts.

11. Using the figures assess fixed assets, current assets and net assets. Net assets should equal your capital + reserves + profit/loss account. You can determine profit (gross and net), and do a simple set of accounts.

    Employ performance and liquidity ratios in your presentation.

12. (a) 6.6%, 22%

    (b) 489 kg, 457 kg, 381 kg

    (c) £1222.50, £1142.50, £952.50

    (d) £80, £270

13. (a) The productive hourly rate = £10.12
    (b) £72.24 (by motorway)
        £64.96 (by road) + 1.5 (hours) + £10.12 = £80.14
    (c) It is most cost effective to go by motorway.
14. (a) The dimensions of the bar are uniform apart from the ends. By taking the average of top and bottom length = 10.1 cm and by adding 1 cm we get a 10% increase in length and weight (±0.1% this is probably within the tolerances of the machine making the bars).
    (b) Cost of new bar = 9.9p
    (c) Cost of promotion = £1 300 000

    Profit on each new bar = 11.1p

    No. of extra bars = 130 000 000/11.1 = 11 711 712

Appendix 1

Standard Deviation

The measure of spread of a series of results is given using standard deviation.

$$\text{standard deviation} = \sqrt{\frac{\Sigma f(x - \bar{x})^2}{N}} = \sqrt{\frac{\Sigma fx^2}{N} - \bar{x}^2}$$

Different symbols are used for standard deviation. The most common ones are s, , sd and SD. The significance of these will become obvious in later studies. At this stage we will not use symbols.

The meaning of this formula and how to use it may be best explained by an example.

Worked example

Six engineering firms paid wages to their employees at the following rates:

	No. of employees	Weekly wage
Firm 1	28	£104
Firm 2	32	£109
Firm 3	16	£112
Firm 4	14	£107
Firm 5	14	£106
Firm 6	54	£108

Determine the mean and standard deviation of these wages.

Answer

We use the formula:

$$\text{standard deviation} = \sqrt{\frac{\Sigma fx^2}{N} - \bar{x}^2}$$

This is the formula we usually use. To do this we must first find $\bar{x}$.

x is the value of each wage.

f is the frequency (number of times) each wage occurs.

$\bar{x}$ is the mean value of the wages.

Therefore for firm 1 $x = £104; f = 28$.

$$
\begin{aligned}
28 \times £104 &= £2\ 912 \\
32 \times £109 &= £3\ 488 \\
16 \times £112 &= £1\ 792 \\
14 \times £107 &= £1\ 498 \\
19 \times £106 &= £2\ 014 \\
54 \times £108 &= £5\ 832 \\
\hline
163 & \quad\quad £17\ 536
\end{aligned}
$$

$$\bar{x} = \frac{\text{total wages}}{\text{total staff}} = \frac{£17\ 536}{163} = £107.58$$

Σ means all the results added up.

N = total number of workmen.

We must now determine $N, \Sigma fx^2$.

f	x	x^2	fx^2	
28	£104	£10 816	£302 848	($28 \times £104^2$)
32	£109	£11 881	£380 192	($32 \times £109^2$)
16	£112	£12 544	£200 704	($16 \times £112^2$)
14	£107	£11 449	£160 286	($14 \times £107^2$)
19	£106	£11 236	£213 484	($19 \times £106^2$)
54	£108	£11 664	£629 856	($54 \times £108^2$)
$N = 163$		$\Sigma fx^2 =$	£1 887 370	

We finally calculate standard deviation.

$$\text{standard deviation} = \sqrt{\frac{£1\,887\,370}{163} - (£107.58)^2}$$

$$= \sqrt{£11\,578.96 - £11\,573.46}$$

$$= \sqrt{£5.50} \quad = £2.35$$

This is very time consuming but most scientific calculators can work out the result from a list of the values given. The program and method of its use are given in the instructions for the calculator.

Can you see that the smaller the value of the standard deviation the closer the values of x are to $\bar{x}$? Therefore standard deviation measures the spread of results.

Appendix 2

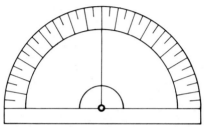

A protractor is used to measure angles, for example, on a pie chart. These are measured off in degrees from the point O. Suppose you want to measure a 50° angle, then a 40° angle. (The symbol ° means degree.)

Draw a circle. Using the centre, place point 0 of the protractor over the centre. Put the base of the protractor along diameter AB. Read off the value 50° on the protractor. Mark point X. Join OX.

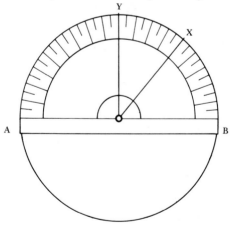

Now 50° + 40° = 90°

Read off the value 90° and mark the point Y. Join OY.

The angle XOB = 50°

The angle YOX = 40°

All angles can be measured in this way.

Index

accounts 117
addition: 8
  decimal fractions 46
  vulgar fractions 30, 40
appreciation 90
approximation 52, 54
area:
  circle 107
  rectangle 105
  triangle 106
arithmetic 7
arithmetic mean 147
assets: 81
  net 81
  current 82
averages: 147
  weighted 151
average return rate 149
axis:
  horizontal 124
  vertical 124

bar charts: 165
  component 168
  horizontal 166
  vertical 165
bid price 79
bills:
  electricity 112, 114
  gas 112, 113
  rates 112
  telephone 115
blending 71
BODMAS 26
break-even chart 131

calculator: 3
  compound interest 87, 89
  notes 25, 57
capacity 62
chain rule 73, 74
charts:
  bar 165
  pie 165

  sales 124
  Z 128
classes 158
clock cards 120
computer 3
costs:
  fixed 131
  variable 131
credit: 82
  ratios 82

debt: 82
  ratios 82
depreciation: 90
  diminishing share balance 91
  equal instalments 90
diameter 109
discounted cash flow 93
discounts: 92
  cash 92
  settlement 92
  trade 92
division: 21
  decimal fractions 48
  long 23, 50
  short 21, 48
  vulgar fractions 38, 42

EFT 4
electronic funds transfer 4

four rules:
  combinations 25
fractions: 30
  decimal 45
    addition 46
    division 48
    multiplication 48
    subtraction 47
  improper 39
  mixed 39, 40
  proper 30, 36, 38

Index

vulgar
  addition 30, 40
  division 38, 42
  multiplication 36, 42
  subtraction 30, 35, 40
frequency distribution 159

graph paper 124
graphs: 124
  straight line 137

histograms 170
historigrams 173

integers 8
interest: 84
  compound 85
  simple 85
invoices 112

length 60
liabilities: 81
  current 82

managed funds 95
margin 98
mark down 98
mark up 98
median 152
metric 60
mixtures 71
mode 152
multiplication: 18
  decimal fractions 48
  long 19
  vulgar fractions 36, 42

offer price 79
overheads 80

percentages 77
perimeter:
  circle 109
  rectangle 108
pictograms 169
pie charts 163
principal 85
profit:
  gross 80

net 80
motive 1
proportion: 69
  direct 70, 137, 138
  inverse 71, 137, 138

radius 107
range 158
ratio 68
ratios:
  liquidity 82
  performance 80
relative frequency percentage 159
repeater 52
return on capital 81
rounding down 54
rounding up 54

salary slip 119
sales 80
sales charts 124
sales volume 181
scrap value 91
shares 96
stock turnover 81
stocks 98
subtraction: 8
  decimal fractions 47
  vulgar fractions 30, 35, 40

tables 155
tally diagrams 159
tolerances 147, 153

values:
  continuous 161
  discrete 161
variables:
  dependent 124
  independent 124
volume: 62
  cuboid 110
  cylinder 110
  sphere 110

wages 119
weight 65

Z charts 128